Environmental Ethics

The Central Issues

Environmental Ethics

The Central Issues

Gregory Bassham

Hackett Publishing Company, Inc.
Indianapolis/Cambridge

Preface

When I first began teaching environmental ethics more than two decades ago, the field was widely seen by professional philosophers as a soft branch of philosophy, mostly taught by left-leaning tree-huggers who lacked the brain power to do any serious work in logic, philosophy of mind, or some other mainstream field of philosophy. Today things are very different. As the global environment has deteriorated and the threat of catastrophic climate change has grown, books, journals, conferences, and college courses in environmental ethics have proliferated, and a great deal of first-rate work has emerged. Now it is widely recognized that deep, rigorous, and creative thinking about the environment is one of the urgent challenges of our time.

Since the first Earth Day (April 22, 1970) more than fifty years ago, environmental awareness, activism, and concern have grown enormously. Yet by most measures of ecological health, our planet is far sicker than it was then. Globally, numerous environmental problems, including biodiversity loss, deforestation, topsoil erosion, ocean acidification, waste disposal, plastic pollution, resource depletion, overpopulation, overconsumption, chemical pollution, overfishing, water contamination, radioactive waste production, and outdoor air pollution, have all become significantly worse. Even more alarmingly, climate change caused mainly by burning fossil fuels has created a dire planetary emergency, threatening to swamp our coastal cities in rising seas; cause massive extinction cascades; create devastating droughts, floods, super-storms, and wildfires; bankrupt our economies; and make large portions of the world uninhabitable by humans.

My goal in this book has been to provide a clear, lively, and balanced introduction to the central issues and controversies in environmental ethics. It is written mainly for students and requires no previous knowledge of philosophy. At the same time, I hope that the book will be of interest to environmental scientists, environmental policy makers, and anyone curious to know what philosophers are saying today about the environment.

The book is divided into two parts. Part One (chapters 1–7) deals with theoretical issues in environmental philosophy. There we examine a variety of ethical and environmental theories that provide concepts, principles, and perspectives that defenders believe should shape our thinking on ecological issues. In Part Two (chapters 8–14), we turn to applied environmental ethics, addressing current debates on topics such as responsibilities to future generations, population growth, overconsumption, food ethics, wilderness preservation, biodiversity loss, climate change, and ecological activism.

I received valuable help in writing this book from Bernard Prusak, Andrew Beck, Jennifer McBride, several anonymous reviewers, and Mia Bassham; my thanks to each. It was a pleasure to work with editors Jeff Dean and Elana Rosenthal at Hackett Publishing; they made this a better book in countless ways. I am grateful to Jim Sterba and Holmes Rolston, who stimulated my interest in environmental ethics when I was a graduate student and newly minted professor. Finally, I thank my students at King's College with whom I explored this subject over many years. If the future is Green, the victory will be theirs.

Part One

Ethics and the Environment

Chapter 1

Ethics: A Brief Introduction

1.1 What Is Ethics?

Environmental ethics is a branch of philosophy that attempts to answer basic questions about how humans should understand and relate to the **environment**.[1] Key questions environmental ethicists ask include:

- What environmental responsibilities, if any, do we have to future generations?
- In thinking about the environment, should we be concerned only with human welfare, or should we also be concerned for the well-being of animals and other life-forms?
- How should we respond to **climate change**?
- Is it ethical to eat animals?
- How important is it to preserve wilderness and areas of great natural beauty?
- How concerned should we be with species extinction?
- Is it ever morally acceptable to break the law to protest some environmental harm? If so, when?

Like business ethics and medical ethics, environmental ethics is a subfield of **ethics**, so let's begin by saying a bit about ethics.

"Ethics" is used in different senses. Sometimes it means "morality," either in the sense of (a) the set of true or correct moral principles (what some people call

1. Although the terms "nature" and "environment" are often used interchangeably, they do not completely coincide. "Nature" generally refers to either (a) the physical universe as a whole or (b) the natural world as it exists apart from human beings and their productions. By contrast, "environment," as that term is used in ecology, refers to the air, water, soils, life-forms, and other external factors that surround and affect a given organism at any time. Thus, although there is only one nature, there are as many environments as there are organisms and time periods. Moreover, many things in nature aren't part of an organism's environment, either because they are too remote to count as part of the organism's "surroundings" or because they have no causal effect on the organism.

"objective morality") or (b) an accepted code of either personal or shared values. Ethics in the sense of accepted norms or values can be studied in a neutral, value-free way. For example, an anthropologist might study the ethical values of a certain tribe in the Amazon jungle, without meaning to endorse those values herself. The study of values in a neutral, objective way is known as **descriptive ethics**. In descriptive ethics, no value judgments are made; that is, nothing is judged to be good or bad, or right or wrong. The goal of descriptive ethics is simply understanding or explanation, not critical appraisal or endorsement.

In this text, we will occasionally delve into issues of descriptive ethics. For example, we will examine how different cultures have thought about nature and the value of nonhuman animals. Mainly, however, we will be engaged in what's called **normative ethics**. In normative ethics, we don't merely study or try to explain codes of conduct; instead, we ask whether they are true, or correct, or rationally defensible. Critical reflection on questions of this sort is the main goal of environmental ethics. When we do environmental ethics, we are not simply learning facts about the environment or studying people's attitudes toward nature. Instead, we are asking about the *best or most defensible* way to think about nature and the amazing variety of living creatures that share our home planet. In short, when we do environmental ethics we are *thinking philosophically* about the environment.

What is **philosophy**? Most simply, philosophy is the attempt to ask and answer fundamental questions about life, value, meaning, and other important issues that cannot be answered by science or empirical observation. Ethics, as we noted, is a branch of philosophy.[2] In ethics, we ask big value questions like these:

- Is there a single correct morality, or is all morality relative to individuals or to societies?
- How should we make moral choices?
- Are there any moral absolutes, or do all moral rules have exceptions?
- What things in life are most worth caring about?
- What kind of person should I be?
- How important is morality? Does it always trump self-interest?

2. In addition to philosophical ethics, there is also theological ethics (also known as moral theology). Theological ethics addresses moral issues from within a particular faith tradition, for example Christianity or Judaism. In this text we will be concerned almost exclusively with philosophical ethics, which does not rely on any religious or faith-based assumptions.

- Does morality depend on religion?
- Is everyone entitled to the same basic rights?
- What does "morally obligated" mean?

Note that there are some important differences between these questions. Some (e.g., "What kind of person should I be?") are normative questions. To answer a normative question, one must make a value judgment about what is good, or right, or valuable, or desirable. Thinkers who attempt to answer substantive questions about what is morally right, wrong, permissible, etc., are engaged in normative ethics. Some ethical questions, however, are *conceptual* rather than normative. For example, the question "What does 'morally obligated' mean?" doesn't require us to make any claims about what is right or good. It's a theoretical question, a query about the meaning of a moral term. When ethicists ask purely conceptual or theoretical questions about morality, they are said to be doing **metaethics** rather than normative ethics. Metaethics deals with the meanings of ethical terms (e.g., "obligation" or "responsibility") as well as with theoretical questions about the nature of ethics and the justification of moral claims. It does not attempt to say what is morally right or wrong, or morally good or bad.

To recap:

- *Normative ethics* attempts to answer questions about what is morally right or wrong.
- *Descriptive ethics* attempts to determine what moral beliefs and values people actually hold.
- *Metaethics* attempts to determine the meanings of moral terms and answer theoretical questions about the nature and foundation of moral values.

1.2 Three Skeptical Views of Ethics

Studies show that many people today—particularly young people—are skeptical of claims that there are "objective" or "absolute" moral truths. Many believe that values are "personal" and that there are no uniquely correct answers to controversial moral issues. Some embrace even deeper forms of skepticism about morality, claiming that there are no "moral truths" or "moral facts" of any sort, perhaps adding that ethics is just a fairy tale that the strong use to dominate the weak. It will be helpful before we begin our study of environmental ethics to explore some of these challenges to traditional morality, because, as we shall see, some of them undermine the whole point of doing environmental ethics.

Moral Subjectivism

Let's begin with the common view that there is no single correct answer to any moral question because morality is entirely "personal" or "subjective." That's what **moral subjectivism** claims. But what does it mean to say that morality is entirely personal or subjective? And is it really true?

Usually when somebody says that ethics is "personal" (or "subjective") they mean that morality is based solely on individual feeling or opinion. On this view, when someone says, for example, that "Abortion is wrong" what they are really saying is something like "I feel that abortion is wrong" or "I believe that abortion is wrong" or "I disapprove of abortion." Because different people obviously have different feelings, beliefs, and attitudes about moral issues, it follows from this view that there is no single correct answer to ethical questions; all morality is "subjective," that is, based solely on personal feelings or opinion.

Subjectivism is attractive to many people today, perhaps because it rules out any kind of judgmentalism (no one can correctly claim that anyone else's moral beliefs are "wrong") and makes everyone moral equals in the sense that no one's ethical beliefs are more correct than anyone else's. A little reflection, however, shows that these aren't genuine attractions at all.

One problem with moral subjectivism is that it implies that there are no real moral disagreements. According to subjectivism, if Sam says, "Affirmative action is right" he's really saying, "I approve of affirmative action." And if Lacey shoots back, "Well, I think affirmative action is wrong," she's saying, "I disapprove of affirmative action." Unless Sam or Lacey is lying about their feelings, they are both saying what is true. But if they are both saying what is true, there is no real conflict between their statements. And if there is no real conflict, there is no real disagreement. But people do clearly disagree about ethical matters. So, subjectivism must be false.[3]

---

3. For a fuller exposition of this objection, see James Rachels and Stuart Rachels, *The Elements of Moral Philosophy*, 5th ed. (New York: McGraw-Hill, 2007), p. 38. Of course, the objection is sound only if it's true that there are genuine moral disagreements. This is something that many subjectivists would deny. Subjectivists can recognize that certain kinds of moral disagreement, such as "disagreements of attitude"—that is, pro- and con-feelings or stances toward things—can be expressed and embodied in moral language. For a lucid discussion of the role of attitudes in moral talk, see Richard B. Brandt, *Ethical Theory* (Englewood Cliffs, NJ: Prentice-Hall, 1959), pp. 206–9.

A second problem with subjectivism is that it implies that no one has any false moral beliefs. As we saw, some people see this as an attraction of subjectivism, because it makes everybody equal in their capacity for moral judgment. According to subjectivism, one person's moral views are just as correct, just as valid, as anybody else's, thus ruling out any kind of ethical elitism or self-righteousness. Such a view certainly has its attractions, but also glaring weaknesses. First, it implies that no one can be morally wiser or more mature than anyone else. A parent, for example, can't claim that they know more about right and wrong than their small child. Second, subjectivism makes it impossible for us to criticize moral views that we find deeply abhorrent. Suppose, for example, that you are strongly opposed to any form of racism. If you are a subjectivist, you can't criticize even the most extreme forms of racism, because, according to your own subjectivist theory, racist views are just as true as your own. Finally, if everyone is always infallible in their moral opinions, there is no point in trying to have *informed* views on moral issues. For example, in deciding what we should do about climate change, simply flipping a coin would be just as effective a way of developing true views as doing research, studying climate science, and so forth. According to subjectivism, no matter how ignorant and uninformed a person may be on an ethical issue, his or her moral views are just as true as anybody else's. For all of these reasons, subjectivism faces serious challenges as an ethical theory.[4]

Moral Relativism

Another popular but flawed ethical theory is **moral relativism**. There are several varieties of moral relativism, but all hold that what is morally right and good varies from individual to individual or from society to society, and thus, there are no "objective" or "absolute" ethical truths. Probably the most widely held form of moral relativism is *cultural moral relativism*, which holds that what is ethically right and good varies from culture to culture. More specifically, cultural moral relativism holds that what is morally right for a person A is what most people in A's culture believe is morally right.

Cultural moral relativism has two main appeals. First, it seems to promote a tolerant, open-minded approach to ethics. It recognizes that moral standards vary from culture to culture, and it seems to rule out any arrogant claims of moral superiority. For example, if you encounter a person from a culture that

4. It should be noted that there are more sophisticated versions of moral subjectivism than the popular version I have criticized. All, in my view, have fatal problems, but some may not be vulnerable to the objections I raise.

practices polygamy, relativism implies that polygamy is morally right in that society, even though you, as an outsider to that culture, may personally strongly disagree. Second, relativism seems to provide an attractive way of dealing with the facts of moral disagreement and the social nature of morality. One of the most striking things about ethics is that people often disagree—sometimes quite vehemently—about moral issues. Relativism explains this by denying the existence of objective or universal moral truths. According to relativists, disagreement is inevitable in ethics because morality is rooted in social convention, not objective moral facts. Lacking any fixed standard for moral judgment, different cultures naturally develop and enforce their own shared moral norms. Each culture tends to regard its own norms as being correct, and relativists agree. On their view, the only true standard of morality is social convention, and this can vary widely from culture to culture. In this way, relativism seems to support the value of tolerance and explain the fact of cross-cultural moral disagreement.

Or so it may seem. In fact, however, cultural moral relativism faces powerful objections. First, it makes it wrong for people to think for themselves about moral issues. Relativism says that what's right and good for you is whatever most people in your society *think* is right and good. Suppose you believe strongly in gender equality but most people in your society hold highly sexist views. In that case, relativism implies that the majority is right and you are wrong. More broadly, relativism implies that the best way to form true moral views is not to gather relevant information, to open-mindedly examine all sides of an issue, and so forth, but simply to consult opinion polls about current moral attitudes. In this way, relativism rules out any kind of critical thinking about moral values.

Second, relativism makes it impossible to object to views that one finds deeply wrong and repugnant. If you strongly oppose human trafficking but most people in your society support it, then you must admit that human trafficking is morally right in your society. Even slavery, child sacrifice, or genocide could, in principle, be defended on relativist grounds.

Finally, relativism can easily lead to inconsistency. Suppose, for example, you are a relativist but most people in your society strongly reject relativism. They believe, for example, that child sacrifice is absolutely and universally wrong. As a relativist, you believe that there are no objective or universal moral truths, but a majority of people in your society disagree. Since relativism says that what is right and good in a society is what most people in that society believe is right and good, it follows that relativism is false in that society. Thus, as a relativist living in a non-relativist society you must both accept relativism and reject it. This, of course, is contradictory.

But what of the alleged attractions of relativism? Doesn't relativism support the important values of tolerance and open-mindedness? And doesn't the fact of deep and pervasive ethical disagreement show that there are no objective moral truths?

Neither of these points, in fact, is a good reason to accept relativism. Consider, first, tolerance. Relativism might support tolerance, or it might not. If you are relativist and live in a tolerant, open-minded society, then relativism implies that you should be tolerant and open-minded. But what if you live in an *intolerant* society? Suppose most people in your society believe they should conquer weaker nations and impose their own values on them? Then you, as a relativist, must embrace intolerance and support the violent subordination of those peoples. Thus, relativism does not necessarily support tolerance, and may in fact compel you—on pain of being inconsistent—to be intolerant.

Finally, what about the fact of deep ethical disagreement? It is true, of course, that there is more debate in ethics than there is in, say, math or science. But this isn't unique to ethics. Consider religion. There, too, we find deep and persistent disagreement. People differ strongly about whether God exists, whether there is one god or many, whether there is an afterlife, and so on. Yet few would claim that truth is relative in religion. As a matter of logic, God either exists or he does not exist. It may be difficult to *know* whether God exists or not. But presumably there is some fact about the matter. Deep and pervasive disagreement does not, therefore, imply that truth is only relative. At most it shows that truth, in that domain at least, is difficult to know.

In short, the apparent attractions of moral relativism don't turn out to be real attractions at all. Given that, plus the serious objections that we noted, relativism should be rejected.

Ethical Egoism

One other popular but seriously flawed ethical theory should be discussed: **ethical egoism**. An egoist is a person who is self-centered, someone who thinks only, or excessively, about himself and his own personal interests. Ethical egoism is a normative ethical theory that asserts that *everyone* should be self-centered. More precisely, *ethical egoism* is the view that each person should pursue his or her own long-term self-interest exclusively, regardless of the consequences to other people. Though such a view seems to negate the whole point of ethics and nearly all moral philosophers reject it, it is surprisingly widely held. Why?

There seem to be several common motivations. Some people are convinced that humans are naturally self-centered, so it's pointless to expect anyone to act unselfishly. Others believe that misguided do-goodism tends to backfire and causes all kinds of social problems, like dependency on government handouts and restrictions on personal freedom. Finally, some people favor ethical egoism because they believe that ethics is a kind of scam or delusion. Such egoists may believe, for example, that in a purely materialistic universe there are no objective moral facts or moral truths, and thus that moral values (unlike self-interest) cannot provide good reasons for action.

Whatever might be said of these various claims, ethical egoism has obvious flaws as an ethical theory. Most obviously, ethical egoism would justify all kinds of illegal and immoral acts. Mass murder, treason, arson, rape, ecocide—any crime on the books could conceivably be justified by appeal to self-interest. In addition, ethical egoism is blatantly arbitrary and self-serving. The ethical egoist claims, in effect, that *his interests*—his pleasures, his comforts, his gratifications—are the only ones that matter. But what makes him so special? Why should his interests trump everybody else's? Simply because they are his? Such a claim fails to treat like cases alike, and so is arbitrary. Finally, ethical egoism fails as an ethical theory because it provides no principled means of resolving human conflicts. If self-interest is the only touchstone of ethical right and wrong, then we will often be stuck with irresolvable conflicts between competing interests. Without ethics, it would be impossible for humans to have strong families and to live together in peaceable, organized societies. Ethical egoism undercuts this by substituting a kind of selfish, dog-eat-dog code of behavior that creates endless conflicts and provides no way of resolving them except in cases of mutual self-interest. In terms of the environment, imagine what our planet would be like if people cared only about themselves and their own personal interests. All the environmental problems we currently face—pollution, resource depletion, overconsumption, climate change, biodiversity loss, and so forth—would immediately become far worse. For all the above reasons, ethical egoism is not an acceptable theory of ethics.

1.3 Three Leading Ethical Theories

Having examined and rejected three common skeptical challenges to traditional ethics—subjectivism, relativism, and egoism—let's now look at three popular and influential ethical theories. Though, as we shall see, none of these theories are problem-free, they do offer useful vocabularies, help to organize our ethical thinking, and draw attention to relevant and important moral considerations. Moreover, as we shall also see, it's helpful to be familiar with these theories because they often play a role in current environmental debates. Let's begin with the simplest and most straightforward of these theories: utilitarianism.

Utilitarianism

Utilitarianism holds that something is morally right if it produces "the greatest good for the greatest number." More precisely, utilitarianism holds that an act is morally right if, and only if, it maximizes overall social benefits, that is, brings about the best possible balance of good welfare-outcomes (for example, happiness or pleasure) over bad welfare-outcomes.

Two English philosophers, Jeremy Bentham (1748–1832) and John Stuart Mill (1806–1873), were the founding fathers of utilitarian moral theory. Bentham and Mill embraced **hedonistic utilitarianism**, which holds that pleasure is the only thing that has value for its own sake. Most contemporary utilitarians deny that pleasure is the sole intrinsic good, and instead embrace some form of **pluralistic utilitarianism** (which claims that many things have **intrinsic value**) or **preference utilitarianism** (which claims, roughly, that an act is right just in case it satisfies the greatest number of desires or preferences). In thinking about happiness or preference satisfaction, some utilitarians claim that only *human* happiness and preferences matter, while others believe that we should also consider the welfare of nonhuman sentient animals, such as primates, dogs, and dolphins.

Applied to the environment, utilitarianism will sometimes produce results that are eco-friendly and sometimes others that are not. On the one hand, many environmental problems (e.g., water pollution or wasteful energy use) would be greatly improved if everybody focused on "the greatest good for the greatest number." On the other hand, focusing solely on utility might mean that turning a pristine mountain into a ski resort would be the right thing to do.

Utilitarianism has long been a popular ethical theory, for reasons that are easy to understand. First, it's a simple theory that often yields clear-cut results. Many ethical theories say that you must consider all kinds of diverse factors (motives, intentions, rules, values, circumstances, past actions, consequences, God's will, etc.) to determine whether an act is morally right or not. By contrast, utilitarians say that you need to consider only *one* thing: the consequences of an act. This greatly simplifies the process of moral choice.

Second, utilitarianism embraces an important kind of equality. Utilitarians believe that everybody's happiness, everybody's interests, *count the same*, regardless of a person's race, gender, wealth, power, or social status. Any form of discrimination that says "My well-being is more important than your well-being" is rejected by utilitarianism.

Finally, utilitarianism appeals to many people because it encourages us to act unselfishly and to strive to build a better world. Lots of problems we face today are due to self-centeredness. Utilitarianism combats this by requiring us to think about everybody's happiness and well-being, not just our own. In this way, utilitarianism seems to have great promise to improve everyday life.

Despite these attractions, critics have raised a number of forceful objections to utilitarianism. For starters, is it really true, as utilitarians claim, that consequences are all that matter in ethics? Suppose five hospital patients will die if they don't receive various organ transplants. Suppose, further, that there's a homeless person (Bill) in the hospital whose tissues are a perfect match for the five patients. The patients will definitely die unless they receive organs from Bill. Suppose, finally, that Bill could be killed secretly, with no chance that anyone would find out. Should Bill be murdered so that five others might live? If consequences are all that matters in ethics, it seems that he should.

What else might matter in ethics besides consequences? What about human dignity and individual rights? Doesn't Bill have a fundamental moral right to life that makes it wrong to murder him in order to harvest his organs? Most of us, I'm sure, would say that he does.

A second common criticism of utilitarianism is that it only considers *net* utility (e.g., net happiness or net welfare), not the *fairness or justice* of how that utility is distributed. Suppose little Bobby absolutely loves chocolate cake. Any time he eats chocolate cake, his happiness level zooms off the charts. Other kids at Bobby's birthday party like chocolate cake but not nearly as much as Bobby does. Would it be right to let Bobby eat *all* the cake if this would maximize net utility? Doesn't the fact that this would be unfair—not to say terribly rude—matter ethically? Not according to utilitarianism.

Finally, is utilitarianism too demanding in what it expects of people? Utilitarianism says we should always do what is *best*—that we always have a moral duty to *maximize* social utility. Think about what that might require. Suppose you love playing in a rock band but realize that you could do much more good in the world if you became a doctor who works in refugee camps. According to utilitarianism, if becoming a doctor and serving in refugee camps would maximize net utility, then it is your moral *duty* to do so. But does morality really require that level of self-sacrifice? Does morality require that we be *saints*—or merely decent and responsible people? Many critics claim that utilitarianism is a kind of "fantasy ethics" because it wrongly demands that we all become ethical saints and full-time utility-maximizers.

Such criticisms of utilitarian ethics might suggest that we need a moral theory that fits better with widely held moral values—what some ethicists call "commonsense morality." That's what defenders of our next ethical theory—duty ethics—try to do.

Duty Ethics

Duty ethics (sometimes called "deontological ethics" or "Kantian ethics") claims that some acts are morally right regardless of their consequences. According to duty ethics, at the heart of morality are moral rules or principles like "Don't steal" and "Tell the truth." Such rules specify ethical duties, like an obligation to keep one's promises or to refrain from causing unnecessary harm. Some of these rules protect individual rights, like the right to own property or to have a fair trial if one is accused of a crime. This is a popular and familiar way of thinking about ethics. Many people, on reflection, would agree that being ethical is at least largely a matter of knowing basic ethical rules, obeying those rules, living up to one's moral duties and responsibilities, and respecting others' rights. The goal of duty ethicists is to defend and systematize such common moral beliefs.

The most famous and influential defender of duty ethics is the great eighteenth-century German thinker Immanuel Kant (1724–1804). Kant tried to work out a theory of ethics that sees morality as grounded, not in good consequences, or social conventions, or feelings, or God's will, but in rationality itself.

Consider the Golden Rule of Jesus: treat others as you would like to be treated. Kant recognized that there's a kind of logical inconsistency in the thinking of unethical people who violate the Golden Rule. A thief, for example, typically thinks it's OK for him to steal from other people, but that it's not OK for other people to steal from him. But what's so special about the thief? Nothing. In morally relevant respects, he's no different from anyone else. Thus, Kant argues, there's a kind of irrationality built into the very idea of immoral behavior.

This insight is the key to what Kant believes is the fundamental principle of morality, which Kant calls the **categorical imperative**. (The principle is "categorical" because it is binding regardless of a person's desires or wishes, and it is "imperative" because it takes the form of a peremptory order or command, rather than a mere suggestion or a piece of advice.) For reasons that aren't entirely clear, Kant formulates the categorical imperative in several different ways, but the best-known formulation—often called the **principle of universal law**—is this:

> *The categorical imperative:* Act only according to that maxim through which you can at the same time will that it become a universal law.

Or more simply: act only in ways that conform to rules that you would like to see adopted by everybody.

Kant points out that a rule might fail to be "universalizable"—that is, capable of, or suitable for, universal adoption—in two different ways. The first is where it's *impossible* for the rule to be universally followed. Consider the rule "Never work; just sponge off others." There's no way everybody could follow this rule, because if nobody worked, there wouldn't be anybody to sponge off of. The second way that a rule could fail to be universalizable is if we ourselves wouldn't *want* it to be universally adopted. Kant gives the example "Never help others in distress." None of us would like to see this universally accepted, because we recognize that *we* might someday be in distress and need some assistance. For these reasons, Kant argues, there's a kind of universality and impartiality built into the very concept of morality. Any kind of special pleading—any claim of the form "I'm special, the ordinary rules don't apply to me"—is not only unethical but irrational.

So how should we decide whether an act is ethical or not? According to Kant, we should ask two questions. First: What "maxim," or general rule, would I be implicitly following if I were to do the act in question? Suppose I'm thinking about cheating on an important test. My maxim in this case would be something like: "It's OK to cheat on an important test." Second: Would I be willing for that rule to be followed by everybody? If the answer is yes, then it would be ethical for me to cheat. If the answer is no, then cheating would be wrong.

Is this a good way to make moral decisions? The key question Kant urges us to ask—"What would happen if everybody did X?"—is certainly relevant in a great many contexts. Kant is right that a good test of moral action should rule out cases of ethical special pleading ("I'm special. It's OK for me to do X but not for anybody else."). But there are problems with Kant's two-step process for making moral decisions.

First, it's not always clear how we are supposed to identify the relevant "maxim." Rules can be formulated broadly or narrowly. Take rules about lying. Kant argues that it is *always* wrong to lie because if everybody lied, there would be no point in lying, because nobody would believe anything anybody said. That's true, but what if we formulated our "maxim" somewhat differently? Suppose a bunch of gang members bang on your door, looking for your teen-aged son, whom they want to murder. Would it be OK to lie and say your son isn't home? Most people would say it is. If so, your operative maxim might be something like: "Always lie when you can save an innocent life by telling a harmless falsehood to a bunch of murderous thugs." That maxim might be universalizable even though the more general rule "Always lie" is not.

Moreover, Kant's universal-law test seems to generate conflicting moral standards. This is because Kant—though he stresses the demands of universal reason—ultimately leaves it up to individuals to decide whether they personally would want a rule to be followed by everybody. Because people's moral intuitions differ, conflicting rules will result from this process. Joe, for example, might be perfectly happy if everybody followed the rule "Always make college admissions decisions race-neutral," while Barbara might strongly oppose such a rule. Thus, Kant's categorical imperative does not produce a single, consistent set of moral principles, as he believed that it would.

Finally, Kant's method for making moral decisions seems to imply that some clearly immoral actions might sometimes be morally permissible. This problem also results from the fact that Kant leaves it up to individuals to determine which rules they would like to see universally adopted. This will vary from person to person, and some people might be willing to universalize obviously immoral maxims. For example, the immoral maxim "Burn all heretics" was widely accepted in the Middle Ages. Kant's two-step universal-law method is faulty because it fails to rule out such clearly bad rules.

Should we, then, reject Kant's categorical imperative entirely? Not necessarily, because as I mentioned, Kant offers more than one formulation of the principle. Many scholars believe that Kant's *second* formulation is much better than his first. This is known as the **principle of dignity**. Kant formulated it as follows:

> *Principle of dignity (second formulation of the categorical imperative)*: Act so that you treat humanity, whether in your own person or in that of another, always as an end and never as a means only.

What did Kant mean by this? And is he right?

Kant believed that there is a huge difference between human beings and mere "things," including all nonhuman animals. Only humans are rational and moral agents. Unlike nonhuman animals, humans can recognize ethical values and freely choose to obey moral law. Only humans can possess what Kant calls a "good will," a desire to do what is right simply because it is right. Kant believed that this gives humans a kind of intrinsic worth or "dignity," that is "above all price." Because humans have this kind of unconditional and transcendent worth, they always deserve respect and should never be treated as a mere "means"—a stepping-stone—to anybody else's pleasures or goals. Humans have a kind of sacredness. They are "ends"—that is, beings with unconditional worth—and must always be treated as such.

This is a powerful and inspiring idea that has spread around the world. It lies at the basis of many systems of ethical and legal thought, including the affirmation in "The United Nations Universal Declaration of Human Rights" (1948) of "the inherent dignity and . . . the equal and inalienable rights of all members of the human family."

Three sorts of criticisms have been leveled at Kant's principle of dignity. The first challenges the reasoning on which it is based, the second claims that the principle is too vague, and the third rejects the notion of "inviolable" dignity.

As we saw, Kant grounds the equal and inherent dignity of human beings on our rationality and our capacity for free and autonomous moral choice. This is what supposedly makes every human being equal in inherent value and equal in their basic moral rights. But is *every* human being rational and capable of moral choice? What of infants, or people with profound mental challenges, or those in a vegetative coma? Here the alleged basis of equal inherent dignity seems to be lacking.

Other critics argue that it's unclear what it means to treat another person as a "mere means." In daily life, we all "use" other people to some extent. For example, we pay people to cook our food, fix our plumbing, and babysit our kids. How can we distinguish between permissible and impermissible ways of using other people to achieve our own goals?

Finally, some critics deny that it is always wrong to treat another person as a mere means. According to these critics, we live in a world where moral trade-offs are unavoidable. Though we might like to believe that all human lives are "beyond price," many people believe that there are tragic circumstances when innocent lives must be sacrificed. In World War II, for example, most Americans supported the use of atomic bombs to speed up the end of the war and save thousands of American lives, even though it was known that this would cause the deaths of tens of thousands of innocent Japanese women, children, and other noncombatants. If this is inconsistent with Kant's principle of dignity, some critics would claim, so much the worse for the principle.

For these and other reasons, few ethicists today agree completely with Kant's duty-centered moral theory. But might some other version of duty ethics work better? One common strategy is to adopt what's called **pluralistic duty ethics**. This claims that, contrary to Kant, there is no single, supreme moral principle, such as the categorical imperative. Instead, there are several basic moral principles that focus on different aspects of proper moral behavior, such as being honest, not harming others, and treating people fairly. The leading defender of pluralistic duty ethics in modern times was the Oxford philosopher W. D. Ross (1877–1971).

Ross argued that ethical conduct is inherently complex and cannot be reduced to any simple decision procedure. He suggests that there are seven main ethical duties. These are:

1. *Fidelity* (i.e., honesty or faithfulness): keeping one's promises and telling the truth.
2. *Reparation*: making up for wrongs we have done to others.
3. *Gratitude*: being thankful for past kindnesses and repaying those who have done us good.
4. *Justice*: treating people fairly and giving them what they deserve.
5. *Beneficence*: doing good to others.
6. *Self-improvement*: striving to become a better person.
7. *Nonmaleficence*: avoiding and preventing harms to others.

None of these duties, Ross believes, is absolute or exceptionless. They can conflict. For example, our duty to tell the truth may sometimes be incompatible with our duty not to hurt other people. For this reason, Ross calls them **prima facie duties** because they can be overridden by weightier ethical considerations and so may hold only "at first glance," or conditionally. When, then, is an act morally right? According to Ross, an act is morally right if and only if it is required by a prima facie duty that is not overridden by any competing and weightier obligation.

Ross's pluralistic brand of duty ethics seems to capture pretty well the way a lot of people think about morality. Many people would agree with Ross that morality, at its core, consists of rules, that many of these rules express duties (e.g., "Don't lie"), that sometimes our moral intuitions pull in different directions, and that ethical choices are not always easy.

That said, there are obvious limitations with Ross's approach to ethics. Most importantly, Ross leaves many important ethical questions unanswered. Why are the seven prima facie duties he identifies the most fundamental ones and not others? And what do we do when one prima facie duty conflicts with another? How should we decide which duty is more important? Ross says we must rely on "intuition"—a kind of direct "seeing" or insight—but what if other people have different intuitions, or our own intuitions are unclear? In short, Ross's pluralistic duty ethics may not be as helpful as we would like a moral theory to be.

Virtue Ethics

The last leading ethical theory we need to consider is **virtue ethics**. This is an old moral theory that goes back to Aristotle and the ancient Greeks. The basic idea of virtue ethics is that the central focus of ethics should be on the questions "What kind of person should I be?" and "What **virtues**, or excellences of character, do I need to be that kind of person?" This is different from many modern moral theories, such as utilitarianism and duty ethics, where the central focus is on the question, "What is the right thing to do?"

Virtue ethics can take different forms. The classic and still most prevalent form of virtue ethics—Aristotle's—is called **eudaimonist virtue ethics**. Aristotle believed that there is a natural goal, or *telos*, of human existence, which he called *eudaimonia* (often translated as "happiness" or "flourishing"). For Aristotle, a *eudaimon*, or flourishing human being, is an all-around excellent person living a good life and engaged in "excellent activities of soul," like high-level thinking and sound moral decision making. Aristotle noted that to flourish in this sense requires a good character, and this, in turn, requires positive character traits, like courage, self-discipline, justice, and sound practical judgment. These are examples of what Aristotle called "virtues." Aristotle discusses two kinds of virtues, intellectual and moral. Intellectual virtues are excellences of the mind, such as good reasoning ability. Moral virtues are habits—ingrained character traits—that persistently incline and empower us to do morally good things. A person has the virtue of honesty, for example, if he or she habitually values and engages in honest behavior. For Aristotle, virtue is essential to human flourishing. We need virtues like courage and moderation to help us achieve "the good life." In fact, Aristotle believed, such virtues are *part* of the good life, because having a good moral character is essential to being an excellent and high-performing person, which in turn is essential to achieving the proper goal of human striving, namely to be a great person living a great human life.

How well does a eudaimonist virtue ethics work as an environmental ethic? Since trashing the environment (e.g., polluting the air and water) would clearly be incompatible with human flourishing, a generally eco-friendly eudaimonist account could easily be fleshed out. But many critics complain that an environmental ethic that focuses solely on *human* flourishing would not be sufficiently nature-friendly. For example, it might allow old-growth forests to be converted to parking lots to satisfy relatively trivial human interests. Some critics also question Aristotle's teleology—specifically, his claim that there is a single natural goal or end that all human beings share. Why think that there is one model of "the good life" that is the same for all people? For these and other reasons, many contemporary virtue ethicists reject Aristotle's eudaimonist (that is, flourishing-based) virtue ethics and prefer other versions of virtue ethics, such as "target-centered" (or "pluralistic") virtue ethics.

Target-centered virtue ethics theories have several components. First, a list of "basic goods" is presented. These might include goods that are necessary for human well-being (e.g., survival, health, knowledge, autonomy, sociability, aesthetic experience, enjoyment, freedom from pain), as well as various non-human-centered goods such as the health and beauty of natural ecosystems and the flourishing of nonhuman forms of life. Second, a theory of virtue is developed based on this list of basic goods. For example, a character trait might be counted as a virtue if it is conducive to promoting one or more of the basic goods. Finally, a target-centered account of right action is offered based on the foregoing theories of basic goods and the virtues. The thought here is that the virtues are aimed at, and serve to promote, various goods. For example, the virtue of friendliness is aimed at the good of sociability, and the virtue of compassion is directed at the good of relieving suffering. An action can then roughly be defined as "morally right" if it adequately (or optimally) "hits the target" of all relevant virtues in a particular moral context.[5]

Target-centered virtue ethics theories can easily be applied to the environment. A clean, healthy environment obviously serves many human and nonhuman basic goods. This allows us to identify various environmental virtues that are targeted at such goods. For example, an environmentally virtuous person would be concerned about unsustainable resource depletion, and so would possess "virtues of sustainability" such as frugality, fairness, and farsightedness. Such a person would also recognize the beauty and intrinsic value of wild places, and so possess "virtues of respect for nature" such as wonder, care, and aesthetic sensitivity.[6] Along such lines a potentially rich and illuminating account might be given of key environmental virtues and values.

Like all ethical theories, environmental virtue ethics has both strengths and weaknesses. With target-centered virtue ethics, the two major concerns are (1) working out a convincing account of human and nonhuman basic goods, and

---

5. This description of target-centered virtue ethics is based mostly on the version of the theory defended in Ronald L. Sandler, *Character and Environment: A Virtue-Oriented Approach to Environmental Ethics* (New York: Columbia University Press, 2007). The leading proponent of target-centered virtue ethics is Christine Swanton. For details on how a target-centered virtue ethics might be developed, see her pioneering book, *Virtue Ethics: A Pluralistic View* (Oxford: Oxford University Press, 2003).

6. Sandler, *Character and Environment*, p. 82.

(2) handling conflicts between competing environmental virtues. On a range of issues, environmental virtue ethics is unlikely to provide clear-cut answers. For instance, would an environmentally virtuous person eat fish? Would they support nuclear power or a substantial reduction in human population? Would they favor immediate, drastic cuts in carbon emissions, even at the cost of dire economic consequences? Like most forms of virtue ethics, environmental virtue ethics tends to deal in such broad generalities that consensus on many important issues is bound to prove elusive. In short, like pluralistic duty ethics, environmental virtue ethics may not be as helpful as we would like a moral theory to be.

1.4 The Value and Limitations of Ethical Theory

In the last section we looked at three leading ethical theories: utilitarianism, duty ethics, and virtue ethics. We saw that each of these theories has both attractions and points of obvious vulnerability. Some of these vulnerabilities, we noted, appear to be serious, if not fatal. So why bother with moral theory? Why not just stick with our everyday, untutored, "common-sense" moral intuitions and standards?

Though no leading ethical theories appear to be fully satisfactory, it is still quite helpful to be familiar with them and to think through their implications on matters of important moral concern. One reason is that "common sense" is not an adequate basis for moral choice. As the great Greek philosopher Socrates (c. 470–399 BCE) pointed out over 2,000 years ago, most of us, when we attempt to think deeply about ethics, turn out to have a pretty confused and inconsistent mishmash of moral beliefs and values. The classical ethical theories can provide a vocabulary and a theoretical framework for deepening and clarifying our moral thinking and making it more coherent. Studying the classical theories can also help us appreciate the depth and complexity of moral problems. This can humble us, motivate us to reflect more deeply, and make us less likely to accept simplistic, pat answers to complicated moral issues. Finally, knowing the classical theories provides an immediate benefit in the study of environmental ethics, because, as we shall see, these theories often play a crucial role in the environmental debates we shall study in this text.

Chapter Summary

1. Environmental ethics is a branch of philosophy that attempts to answer basic questions about how humans should understand and relate to the environment. Environmental ethics addresses such questions as: What environmental responsibilities, if any, do we have to future generations? Is it ethical to eat animals? How concerned should we be with species extinction? and How should we respond to climate change?
2. Three skeptical views of traditional ethics were examined: moral subjectivism (the view that ethics depends on personal opinion), moral relativism (the view that ethics depends on popular opinion), and ethical egoism (the view that each person should be concerned solely with his or her own long-term self-interest). Each of these views, we noted, is open to serious objections.
3. Many environmental ethicists believe that general ethical theories can be of significant value in addressing environmental problems and policies. Three leading ethical theories were examined: utilitarianism, duty ethics, and virtue ethics. Utilitarianism is the view that an act is morally right if, and only if, it maximizes net social utility, that is, produces the best possible balance of good welfare-outcomes (e.g., happiness or preference-satisfaction) over bad welfare-outcomes. Duty ethics claims that ethics is fundamentally a matter of following correct moral rules and performing one's ethical duties. Virtue ethics contends that ethics is not centrally a matter of rules, duties, or consequences, but rather of character and the virtues. We saw that each of these theories has both strengths and weaknesses. Though none of the theories seem fully adequate, it is still helpful to be familiar with them in the study of environmental ethics.

Discussion Questions

1. What is moral subjectivism? Is it a sound ethical theory? Why or why not? In general, is moral subjectivism an eco-friendly ethical theory or not?
2. What is moral relativism? Is it a sound ethical theory? Why or why not? In general, is moral relativism an eco-friendly ethical theory or not?

3. What is ethical egoism? What are its strengths and weaknesses as an ethical theory? In general, is ethical egoism an eco-friendly ethical theory or not?
4. What is utilitarianism? Is it a sound ethical theory or not? How, in general, would a utilitarian treat the environment?
5. What is duty ethics? How does Kant's duty ethics differ from pluralistic duty ethics? Which version of duty ethics is better? How, in general, would an advocate of duty ethics treat the environment?
6. What is virtue ethics? How does eudaimonist virtue ethics differ from target-centered virtue ethics? Which version of virtue ethics is better? How, in general, would an advocate of virtue ethics treat the environment?

Further Reading

The free, online *Stanford Encyclopedia of Philosophy* contains authoritative, up-to-date (but sometimes rather advanced) discussions of all the ethical theories discussed in this chapter. In *The Elements of Moral Philosophy* (McGraw-Hill, 9th ed., 2018), James and Stuart Rachels offer a clear and highly readable introduction to ethics aimed at beginning philosophy students. For utilitarianism, see John Stuart Mill's *Utilitarianism* (originally published in 1863 and now free online). For Kant's moral philosophy, see his classic *Groundwork of the Metaphysics of Morals*, translated by Mary Gregor and Jens Timmermann (Cambridge University Press, 2012). For virtue ethics, see Aristotle, *Nicomachean Ethics*, 3rd ed., translated by Terence Irwin (Hackett, 2019) and Rosalind Hursthouse, *On Virtue Ethics*, rev. ed. (Oxford University Press, 2002).

Chapter 2

Religion and the Environment

Religion has long been, and remains, a powerful influence on how people think about nature and the environment. All of the great world religions include teachings that speak to ecological concerns. In this chapter we'll examine some important religious perspectives on the environment. Let's begin with Christianity and then turn to consider some non-Western and Native American views.

2.1 Christianity and the Environment

Has Christianity generally been an eco-friendly religion? Not according to the distinguished medieval historian Lynn White (1907–1987). In 1967, White published a short article titled "The Historical Roots of Our Ecologic Crisis" that set off a firestorm of controversy. In that essay White argued that Christianity is largely to blame for the world's environmental problems because it teaches a view of humanity's place in nature that encourages environmental domination and destruction.

White draws attention to passages in Genesis, the first book of the Bible, that seem to teach strongly human-centered views of nature. There, humans are said to have been uniquely created in the "image" and "likeness" (Gen. 1:26) of God, making them superior to all other forms of life on earth. In the first chapter of Genesis, God commands the earliest humans, Adam and Eve, to be "fruitful and multiply, and fill the earth and subdue it; and have dominion over the fish of the sea and over the birds of the air and over every living thing that moves upon the earth" (Gen. 1:28). Later, after the great flood, God repeats his commandment to be fruitful and multiply to Noah and his sons, adding: "The fear of you and the dread of you shall be upon every beast of the earth, and upon every bird of the air, upon everything that creeps on the ground and all the fish of the sea; into your hand they are delivered" (Gen. 9:1–2). According to White, for many centuries such passages were widely interpreted by Christian thinkers to teach that everything in nature was created solely for human benefit, making

Christianity "the most **anthropocentric** religion the world has seen."[1] Following the Scientific Revolution, White claimed, humans used their new scientific and technological know-how to cut down forests, pollute rivers and the air, poison the land, and drive many animals to extinction. The only solution to the environmental crisis, he argued, lies in rejecting the biblical conquer-and-subdue mindset. Ultimately, he said, we must "find a new religion, or rethink our old one."[2] As a church-going Christian himself, White strongly preferred the second option, and proposed Saint Francis of Assisi (1181/1182–1226) as the patron saint of **ecology**. Francis preached to birds and sang hymns to "Brother Sun," "Sister Mother Earth," Brother Wind," and "Sister Water." According to White, Francis embraced an unorthodox "pan-psychic" view of nature, seeing all things as alive and deserving of respect. While such an animistic view may seem more pagan than Christian, White agreed with Saint Francis's fundamental insight that we "must rethink and re-feel our nature and destiny,"[3] because no long-term solution to the ecological crisis is possible as long as people continue to embrace "orthodox Christian arrogance toward nature."[4]

White's provocative article set off a heated debate, with both defenders and critics jumping into the fray.[5] Many critics complained that White overstated the role of religion in causing environmental problems. It was pointed out, for example, that many non-Christian cultures, such as China, India, and Japan, also have rather poor environmental records,[6] and that Greek and Roman

---

1. Lynn White, Jr., "The Historical Roots of Our Ecologic Crisis," *Science*, vol. 155 (March 10, 1967), pp. 1203–7; reprinted in J. Baird Callicott and Robert Frodeman, eds., *Encyclopedia of Environmental Ethics and Philosophy*, vol. 2 (Farmington Hills, MI: Macmillan Reference, 2009), p. 443.

2. White, "Historical Roots," p. 444.

3. White, "Historical Roots," p. 445.

4. White, "Historical Roots," p. 445.

5. Some defenders of White have sought to bolster his brief argument with additional evidence. For instance, environmental historian Roderick Frazier Nash has drawn attention to the "pervasive otherworldliness" of Christianity, the traditional Christian view of wilderness as a hostile and cursed land, the Christian idea that God will destroy the earth at the last trump, and the fact that the Hebrew words for "subdue" (*kabash*) and "have dominion over" (*radah*) are used throughout the Old Testament to signify a violent assault or crushing. Nash, *The Rights of Nature: A History of Environmental Ethics* (Madison, WI: University of Wisconsin Press, 1989), pp. 90–91.

6. Lewis W. Moncrief, "The Cultural Basis for Our Environmental Crisis," *Science*, vol. 170 (October 30, 1970), pp. 508–12; J. Patrick Dobel, "Stewards of the Earth's Resources: A Christian Response to Ecology," *Christian Century* (October 12, 1977), pp. 906–9; reprinted under the title "The Judeo-Christian Stewardship Attitude to Nature," in Louis P. Pojman and Paul Pojman, eds., *Environmental Ethics: Readings in Theory and Application*, 6th ed. (Boston, MA: Wadsworth, 2012), pp. 628–32.

thought also significantly contributed to Western anthropocentric attitudes.[7] Another common complaint was that White offered a selective and one-sided reading of the Bible. Many Christian critics argued that White overlooked passages in Scripture that indicated that humans should be good "stewards" of nature, rather than conquerors and destroyers.[8] Let's take a closer look at this important notion of **environmental stewardship**.

2.2 Environmental Stewardship

"Steward" is an Old English word, meaning, most commonly, "one who manages the affairs of an estate on behalf of an employer." As applied to the environment, a steward would thus be someone who is appointed to manage and care for the earth and its creatures on behalf of its rightful owner, God. A good steward, on this view, would be a kind of faithful trustee, someone who cares for, preserves, and perhaps improves the earth and its ecosystems in a way that God would approve. According to many critics of White, when the Bible speaks of "dominion" it means faithful stewardship, not domination or conquest. How convincing is this claim?

The Bible teaches that God created the world and is its true owner and ruler.[9] At the same time, God has in some sense entrusted the earth to humans, given us "dominion" over it.[10] But this dominion is far from absolute; God

7. John Passmore, *Man's Responsibility for Nature: Ecological Problems and Western Traditions* (New York: Charles Scribner's Sons, 1975), p. 13.

8. As White later noted, this criticism misses the mark, because his point was about how the Bible has traditionally been interpreted and not how it should properly be interpreted. See Lynn White, Jr., "Continuing the Conversation," in Ian G. Barbour, ed., *Western Man and Environmental Ethics: Attitudes toward Nature and Technology* (Reading, MA: Addison-Wesley, 1973), pp. 60–61.

9. Ps. 24:1 ("The earth is the Lord's and the fulness thereof"); Ps. 89:11–12 ("The heavens are thine, the earth also is thine, the world and all that is in it"); Jer. 27:5 ("It is I who by my great power and my outstretched arm have made the earth, with the men and animals that are on the earth, and I give it to whomever it seems right to me"); Ps. 50:10–11 ("For every beast of the forest is mine, the cattle on a thousand hills. I know all the birds of the air, and all that moves in the field is mine"); and Lev. 25:23 ("The land shall not be sold in perpetuity, for the land is mine; for you are strangers and sojourners with me"). All quotations from the Bible are from the Revised Standard Version.

10. Gen. 1:26 ("The God said, 'Let us make man in our image, after our likeness; and let them have dominion over the fish of the sea, and over the birds of the air, and over the cattle, and over every creeping thing that creeps upon the earth.'"); Gen. 9:3 ("Every moving thing that lives shall be food for you; and as I gave you the green plants, I give you everything"); Ps. 8:6 ("Thou hast given him dominion over the works of thy hands;

remains the ultimate ruler of his creation, and he imposes many limits on what humans can rightfully do to nature. For example, Adam and Eve were commanded to "till" and "keep"[11] the Garden of Eden, and were originally permitted to eat only plants.[12] God also commands the Jews to observe various dietary restrictions,[13] to have "regard" for the life of animals,[14] not to muzzle an ox when it is treading out the grain,[15] to let fields lie fallow every seventh year so that the poor and wild animals may eat,[16] and not to cut down food-bearing trees when besieging towns.[17] Moreover, as the handiwork of God, nature praises and glorifies its Creator,[18] manifests his wisdom and power,[19] and is regarded by God as being "very good."[20] It would seem to follow from such passages that any wanton destruction of nature or waste of natural resources is contrary to the divine will, and that we should instead be conscientious stewards of the natural gifts and resources that God has entrusted to our care.

Over the past few decades, stewardship has become the reigning paradigm in Judeo-Christian thought.[21] But there are difficulties with the notion of stewardship that should not be ignored.

For starters, of course, the whole idea of stewardship presupposes a set of religious beliefs (e.g., that God exists, that he "owns" the created universe, that he has given humans "dominion" over the earth, and that we can at least

---

thou hast put all things under his feet"); and Ps. 115:16 ("The heavens are the Lord's heavens, but the earth he has given to the sons of men").

11. Gen. 2:15.

12. Gen. 1:29. Even animals, at this stage, ate a strictly vegetarian diet (Gen. 1:30). Meat eating was permitted only after the Great Flood (Gen. 9:2–3). According to Isaiah (11:7), animals will no longer be carnivorous in the Messianic Age ("the lion shall eat straw like the ox").

13. For example, Deut. 14:3ff.

14. Prov. 12:10.

15. Deut. 25:4.

16. Exod. 23:10.

17. Deut. 20:19.

18. For example, Ps. 148:3–10.

19. Ps. 19:1 ("The heavens are telling the glory of God; and the firmament proclaims his handiwork").

20. Gen. 1:31. Cf. 1 Tim. 4:4 ("Everything created by God is good"); and Wis. 11:24 ("For you love all things that exist, and detest none of the things that you have made"). It should be noted that when God declared creation to be "very good" he was speaking of pre-fallen nature. By the time of Noah, the world had become so wicked that he was "sorry" that he had made humans and other animals (Gen. 5:7). Nevertheless, mainstream Christian thought has always regarded the world as being fundamentally good.

21. Robin Attfield, "Christianity," in Dale Jamieson, ed., *A Companion to Environmental Philosophy* (Malden, MA: Blackwell, 2001), pp. 107–9; and Nash, *The Rights of Nature*, pp. 95–113.

roughly know how he wants us to exercise that dominion) that many people would reject. Pointing to its religious underpinnings, the late Harvard scientist and agnostic Stephen Jay Gould (1941–2002) rejected the whole concept of environmental stewardship. He writes:

> Such views, however well intentioned, are rooted in the old sin of pride and exaggerated self-importance. We are one among millions of species, stewards of nothing. By what argument could we, arising just a geological microsecond ago, become responsible for the affairs of a world 4.5 billion years old, teeming with life that has been evolving and diversifying for at least three-quarters of that immense span? Nature does not exist for us, had no idea we were coming, and doesn't give a damn about us.[22]

Moreover, as Gould suggests, most conceptions of stewardship have an anthropocentric bent that many environmentalists believe to be harmful and unwarranted. As we saw, the Bible seems to imply that God has commanded humans to "fill the earth and subdue it" (Gen. 1:28) because they alone were created in the image and likeness of God and thus are superior to all other earthly life-forms. Many environmentalists today believe that such views are arrogant and self-serving.

Many critics would also question whether humans possess the wisdom and ability to be effective stewards of nature. Natural processes are often incredibly complex, and history has repeatedly shown that when humans try to "manage" nature we frequently end up doing more harm than good. Given the dismal human track record on the environment, one can question whether an all-wise and providential God would have placed nature under our dubious care.

Finally, the idea of stewardship is quite vague and open-ended. A good steward would be one who manages the earth and its natural resources in a way that God would approve. But how can we know what environmental actions God favors? How does God feel about nuclear power, or fracking, or reducing world population, or damming scenic rivers, or genetically modified foods, or trapping wild animals, or driving gas-guzzling pickup trucks, or zoos, or restoring wolves to the Great Plains, or building oil pipelines in the Arctic? Asking what God thinks about such issues is largely a matter of guesswork. As a result, any detailed model of environmental stewardship would inevitably be controversial and rest on highly debatable theological premises.

---

22. Stephen Jay Gould, "The Golden Rule—A Proper Scale for Our Environmental Crisis," *Natural History* 90:9 (September 1990), pp. 24–30; reprinted in Stephen Jay Gould, *Eight Little Piggies: Reflections in Natural History* (New York: W. W. Norton, 1994), p. 48.

MAKING A DIFFERENCE

Pope Francis: The Environmental Pope

In 2015, Pope Francis published *Laudato Si'*, a landmark papal encyclical on the environment. In that work, Pope Francis broke sharply with the Church's traditional strongly human-centered approach to nature and issued a bold call for an "ecological conversion" to address urgent environmental concerns, including climate change, pollution, environmental injustice, access to clean drinking water, unsustainable patterns of **consumption**, and the rapid global collapse of **biodiversity**.

Traditionally, Catholic popes and theologians have paid little heed to ecology, and most have adopted what Pope Francis calls "extreme" anthropocentric views of creation that see humans as separate from nature, superior to nature, and authorized by God to have a kind of tyrannical "dominion" over the rest of creation. In *Laudato Si'*, Pope Francis offers a strikingly different vision of how humans should understand their place in nature—one rooted in the "fraternalist" vision of creation espoused by the pope's namesake, Saint Francis of Assisi (c. 1181/1182–1226). Instead of thinking of ourselves as masters and conquerors of the natural world, the pope urges that we see earth as the "common home" we share with our nonhuman "family" and "community" of coevolved life-forms. Though humans do indeed have a special kind of dignity and were given the earth as a common gift from God, we should not confuse "dominion" with subjugation. Instead, we should recognize that God loves all creatures; that all living things have value in themselves; that plants and animals were not created solely for human benefit; that all life-forms on earth are part of the same biotic family; and that humans were intended to be faithful trustees of earth's natural treasures, not ruthless exploiters of them. In short, in *Laudato Si'* Pope Francis sought to update church teaching about the environment by substituting an ecologically informed "good stewardship" model of environmental responsibility for the markedly human-centered views that had long prevailed in Catholic thought.

2.3 Non-Western and Native American Attitudes toward Nature

As we've seen, Christianity has a quite checkered record on the environment. How do other religious traditions stack up? Let's briefly examine three Eastern religions or spiritual traditions—Buddhism, Taoism, and Hinduism—as well as traditional Native American thought.

Buddhism

The world's fourth largest religion, Buddhism originated in India sometime around the fifth century BCE and later spread to China, Japan, Thailand, and much of Asia. Believing that life is full of suffering and unsatisfied desire, Buddhists seek to achieve inner peace, enlightenment, freedom from suffering, and liberation from the cycle of death and rebirth (*samsara*) by means of meditation and disciplined control of one's desires. Buddhists reject strongly human-centered views of the environment. The founder of Buddhism, Siddhartha Gautama (c. 490–c. 410 BCE), forbade his fellow monks from practicing agriculture or traveling during the rainy season because they might step on worms, insects, and other forms of life. The first of Buddha's Five Precepts is to refrain from harming sentient life. Many Buddhists believe that all sentient things have a "Buddha-nature" that entitles them to compassion and respect. Core Buddhist values include nonviolence (*ahimsa*); compassion (*karuna*); loving-kindness (*metta*); recognition of the causal interconnectedness of all things; and a rejection of materialism, egoism, and greed. Buddhists believe that loving-kindness should be extended not only to our fellow humans but to all suffering creatures. Speaking to his disciples, Buddha said, "As a mother, even at the risk of her own life, protects her son, her only son, so let him cultivate love without measure toward all beings. Let him cultivate toward the whole world—above, below, around—a heart of love unstinted, unmixed with the sense of differing or opposing interests."[23] Many Buddhists are vegetarians, in large part because of their belief in human and animal reincarnation.

Taoism

Taoism (or Daoism) is an ancient Chinese religion or wisdom tradition that stresses simplicity, humility, balance, and harmony with nature. According to

---

23. Quoted in John B. Noss, *Man's Religions*, 5th ed. (New York: Macmillan, 1974), pp. 135–36.

Chinese tradition, Taoism was founded by Lao Tzu (sometimes spelled Laozi, meaning "Old Master"), who lived in the 6th century BCE. Lao Tzu reputedly wrote a short book, the *Tao Te Ching* ("The Book of the Way and Its Power," now often spelled *Daodejing*), which is the fundamental text of the Taoist tradition. The central concept in Taoism is that of the Tao, which means "Way" or "Path." The Tao is simultaneously the way of Ultimate Reality, of Nature, and the way humans should live their lives. Though the Tao cannot be grasped through words or rational thought, it can be sensed through intuition in moments of quietude and tranquility. That is why Taoists have always valued peaceful forests and scenic areas of unspoiled nature. They view nature as a harmony of opposing forces (hot and cold, light and dark, etc.) and seek to live simple, contented lives in tune with nature. Believing that "a good traveler leaves no track or trace,"[24] they seek to "embrace simplicity, reduce selfishness, [and] have few desires."[25] To symbolize their commitment to harmony with nature, Taoists usually build their temples nestled against hills, or shaded by leafy trees, so that they blend into their natural surroundings.[26] The central Taoist idea of *wu-wei* (literally "non-doing") emphasizes the value of "letting things happen naturally" and without meddlesome interference, an idea that resonates with much modern environmental thought.

Hinduism

Like Buddhism, Hinduism is an ancient religion that originated in India. Though there are many schools of thought in Hinduism, most believe in certain core doctrines. One is the ultimate Oneness of reality. For Hindus, there is no sharp separation between God and nature, or between humans and nonhuman animals, because all things are ultimately aspects or manifestations of one Supreme Reality. Such a view readily lends itself to attitudes of respect or reverence for nature.

Like Buddhists, Hindus also believe in reincarnation. Certain animals, such as cows, are sacred to Hindus because they believe that various Hindu deities were once incarnated in those animal forms. Hindus also believe that human beings are sometimes reborn as animals. This is one reason many Hindus are vegetarians and practice *ahimsa* (nonviolence and avoidance of injury).

24. *Tao Te Ching*, in Wing-Tsit Chan, ed., *A Source Book in Chinese Philosophy* (Princeton, NJ: Princeton University Press, 1963), p. 153.
25. *Tao Te Ching*, p. 149.
26. Huston Smith, *The Religions of Man* (New York: Harper & Row, 1986), p. 284.

Traditionally, many Hindus have practiced a form of tree worship. Believing that trees are conscious and symbolize various aspects of God, they recognize religious duties to plant, protect, and reverence trees and other plants.[27]

Finally, many sacred Hindu texts condemn materialism and greed, and affirm the values of moderation, nonattachment, and self-control. This is reflected in the Hindu doctrine of the Four Stages of Life, where the final stage is that of a forest-dweller (*sannyasi*). In this period, elderly men and women retire from the world of work and family and live simple, austere lives of renunciation, meditation, and purification.[28]

Traditional Native American Religion

Native American attitudes toward nature are varied and diverse, but as environmental philosopher J. Baird Callicott notes,[29] there are some widely shared themes. These include belief in: a Great Spirit, the Father and creator of all living and nonliving things, often identified with the Sky; an Earth Mother, who together with the Great Spirit gives life and connectedness to all earth-dwelling beings; pan-psychism, the idea that all things—even material objects like stones or mountains—are alive and endowed with qualities of personhood; holy places—sacred locales of great spiritual potency; the importance of belonging to, conserving, and caring for the land and natural resources; the impossibility of "owning" the land in the proprietary European sense; and the universal kinship of all earthly creatures, who are all equally children of the Great Spirit and endowed with a spark of his divinity.

Environmentalist Karen Warren reports a conversation she had with a Sioux elder[30] that nicely illustrates the Native American attitude toward nonhuman creatures. One day the elder sent his seven-year-old son to a reservation to live with his grandparents so that he could learn traditional Indian ways.

---

27. O. P. Dwivedi, "*Satyagraha* for Conservation: A Hindu View," in J. Ronald Engel and Joan Gibb Engel, eds., *Ethics of Environment and Development: Global Challenge, International Response* (London: Bellhaven Press, 1990); reprinted in Louis P. Pojman and Paul Pojman, eds., *Environmental Ethics*, pp. 644–45.

28. Noss, *Man's Religions*, p. 191.

29. J. Baird Callicott, "Traditional American Indian and Western European Attitudes toward Nature: An Overview," *Environmental Ethics* 4 (1982), pp. 293–318; reprinted in Frederick A. Kaufman, ed., *Foundations of Environmental Philosophy: A Text with Readings* (New York: McGraw-Hill, 2003), p. 65.

30. Warren doesn't mention the ethnic group of the Sioux elder, whether Dakota, Yankton, or some other tribal grouping.

There the boy's grandparents taught him how to hunt the "four leggeds" of the forest. The boy was instructed

> to shoot your four-legged brother in his hind area, slowing it down but not killing it. Then, take the four legged's head in your hands, and look into his eyes. The eyes are where all the suffering is. Look into your brother's eyes and feel his pain. Then, take your knife and cut the four-legged under his chin, here, on the neck, so that he dies quickly. And as you do, ask your brother, the four-legged, for forgiveness for what you do. Offer also a prayer of thanks to your four-legged kin for offering his body to you just now, when you need food to eat and clothing to wear. And promise the four-legged that you will put yourself back into the earth when you die, to become nourishment for the earth, and for the sister flowers, and for the brother deer.[31]

Recent studies have shown that Native Americans were not always as earth friendly as has sometimes been assumed. When European settlers first reached North America, they encountered a landscape that was already extensively altered by human activity.[32] Native peoples frequently used controlled (and not so controlled) burns to facilitate hunting and growing crops. Their use of fire and hunting techniques, like driving herds of buffalo off cliffs, often resulted in wasteful mass kills. George Catlin, who admired Native Americans and lived with them for many years, wrote of the "improvident character" of the Sioux Indians, who in one afternoon slaughtered 1,400 buffalo simply for their tongues, leaving the carcasses to rot or be devoured by wolves.[33] Still, there is no question that Native Americans generally had—and have—great respect for nature.

---

31. Karen J. Warren, "The Power and Promise of Ecological Feminism," *Environmental Ethics* 12:3 (1990), pp. 125–46; reprinted in Kaufman, ed., *Foundations of Environmental Philosophy,* pp. 427–28. Those who have seen James Cameron's top-grossing 2009 film *Avatar* will recall that the indigenous humanoids in that film hold very similar attitudes toward their world's "four leggeds."

32. William Cronon, *Changes in the Land: Indians, Colonists, and the Ecology of New England* (New York: Hill and Wang, 2003).

33. George Catlin, *Letters and Notes on the Manners, Customs, and Conditions of the North American Indians, Written during Eight Years' Travel Amongst the Wildest Tribes of Indians in North America* (New York: Wiley and Putnam, 1841); reprinted in Robert Finch and John Elder, eds., *The Norton Book of Nature Writing* (New York: W. W. Norton, 1990), p. 141.

2.4 Conclusion

Throughout history, religion has had a major impact on how people have viewed nature. As we have seen, religions have had a mixed record on the environment. As historian Patrick Dobel has argued, "all cultures, regardless of religion, have abused or destroyed large areas of the world either because of economic or population pressures or from simple ignorance."[34] But some religions have tended to be more environmentally friendly than others. Some Eastern religions, such as Buddhism, Hinduism, and Taoism, have generally encouraged attitudes of respect for nature and nonhuman animals. This is also true of Native American spirituality. The Judeo-Christian tradition has a more problematic record on the environment and the treatment of nonhuman animals. Critics, such as the historian Lynn White, have argued that in the Christian West, certain passages in the Bible were widely interpreted as encouraging strongly human-centered attitudes toward nature—attitudes that have greatly contributed to environmental degradation. In recent decades, many Jewish and Christian scholars have argued that such readings were mistaken, that Scripture as a whole teaches that humans should seek to be good stewards of the environment. Indeed, throughout the world, the "greening of religion" is a significant development in the modern environmental movement. Given the importance of religion in shaping people's attitudes toward nature, the growing alliance between religion and environmentalism must be seen as good news for the planet.

Chapter Summary

1. Religion has long been, and remains, a powerful influence on how people think about nature and the environment. All of the world's great religions include teachings that speak to ecological concerns.
2. Some religions have better environmental records than others. Historically, the Judeo-Christian tradition cannot be given high marks. As historian Lynn White argued in a seminal 1967 article, Christian thought has tended to be strongly human-centered. The Bible speaks of God giving humans "dominion" over nature, and this was long interpreted as conferring a right of domination or despotic control over earth

34. Dobel, "The Judeo-Christian Stewardship Attitude to Nature," p. 630.

and nonhuman nature. In recent decades, many Jewish and Christian scholars have argued against this traditional reading of the Bible. They argue that "dominion" should not be understood as domination but rather as faithful stewardship of nature's beauty and resources.

3. The concept of religiously based environmental stewardship is not without its problems. Some critics see it as arrogant and dangerous for humans to believe that they are superior beings who have been put in charge of nature. In addition, being a good steward seems to require that we treat nature as God desires or commands, but it is often unclear how we can know what those desires or commands are.

4. Some major world religions have comparatively good environmental records. This is generally true of Buddhism, Taoism, and Hinduism, all of which emphasize the essential harmony or oneness of humans with nonhuman nature. Respect for nature and the relatedness of all life-forms is also a prominent theme in many strands of Native American spirituality.

Discussion Questions

1. Is Christianity, on the whole, an eco-friendly religion or not?
2. Should humans see themselves as divinely appointed environmental stewards? If so, how generally would good stewards treat the environment?
3. Should humans view themselves as above nature or as part of nature?
4. Do you agree that Buddhism, Taoism, and Hinduism are generally eco-friendly religions?
5. To what extent should we embrace the Taoist notion of "letting things happen naturally" in our treatment of the environment?
6. Should we accept pan-psychism—the belief that all things in nature are alive and sentient? If so, why?

Further Reading

For a helpful overview of religious perspectives on the environment, see John Hart, ed., *The Wiley Blackwell Companion to Religion and Ecology* (Hoboken, NJ: Wiley Blackwell, 2017). Also worth consulting is Richard

Bohannon, ed., *Religions and Environments: A Reader in Religion, Nature, and Ecology* (New York: Bloomsbury Academic, 2014). Lynn White's classic essay, "The Historical Roots of Our Ecologic Crisis" (*Science*, vol. 155, March 10, 1967, pp. 1203–7), has been widely anthologized and is still very much worth reading. On Native American spirituality and the environment, see Bruce Morito, "Native Americans," in J. Baird Callicott and Robert Frodeman, eds., *Encyclopedia of Environmental Ethics and Philosophy*, vol. 2 (Farmington Hills, MI: Macmillan Reference, 2009), pp. 85–90.

Chapter 3
Animal Rights

People often have conflicting feelings about nonhuman animals. On the one hand, humans eat animals, hunt them, trap them, lock them in cages, experiment upon them, train them to perform circus tricks, and exploit them in countless other ways. On the other, many of us have pets that we cherish and treat as full-fledged members of our families. For most of history, humans have generally treated animals as mere resources or commodities, lacking in all moral value. Is this ethical? Or do animals have certain rights (for example, a right not to be killed or treated cruelly) that we should respect? In this chapter we'll look at the arguments for and against animal rights.

3.1 The Case for Animal Rights

There are two broad strategies for defending animal rights. One is to draw attention to the needless suffering that results when humans exploit animals for our own benefit. The other is to argue that some animals possess inherent value and deserve respect for many of the reasons that humans do. The first strategy is a kind of utilitarian argument for animal rights, and the second is based on a duty-centered approach to ethics. Let's begin with the utilitarian argument.

The Utilitarian Argument for Animal Rights

The best-known defender of a utilitarian approach to animal rights is the Australian philosopher Peter Singer (1946–). In his classic 1975 book, *Animal Liberation: A New Ethics for Our Treatment of Animals*,[1] which some refer to as

1. Peter Singer, *Animal Liberation: A New Ethics for Our Treatment of Animals* (New York: New York Review of Books, 1975); 2nd ed. (1990). An updated edition with a new Preface and new subtitle was published by HarperCollins in 2009.

"the Bible of the animal rights movement,"[2] Singer argues for the "liberation" of animals from human "tyranny." The core of his argument can be summarized as follows.

Many animals, like humans, are conscious, sentient beings that can experience suffering and enjoyment. Any being that is sentient has "interests," that is, matters that they have a stake in, things that can make its life better or worse. A dog, for example, has an interest in not suffering hunger or thirst. Human beings, of course, have many interests that animals do not. For example, only humans have an interest in being taught to read or having a right to vote. But humans and some animals share certain common interests. For instance, a human child and a kitten both have an interest in avoiding a broken leg. Morality requires that similar interests be treated similarly unless there is good reason not to do so. (Singer calls this the **principle of equal consideration of interests** or, more simply, "the principle of equality.") Thus, the interests of every sentient being affected by an action should be considered and given the same weight as the like interests of any other sentient being. Humans violate this principle and are guilty of **speciesism** (that is, unjust bias in favor of one's own species) when they raise animals for food, experiment upon them, put them in zoo cages, or engage in other practices that systematically favor human interests over the like interests of animals. It is wrong to cause suffering without adequate justification, and humans have no adequate justification for much of the suffering they inflict on animals. Thus, humans should fundamentally rethink the way they treat animals and stop causing them needless suffering.

This is a powerful argument that has convinced many people to reexamine their attitudes toward animals. However, the argument also has some points that can be questioned. Let's examine a few.

First, it is unclear what practical implications flow from Singer's argument. Few would disagree with the virtual truism that equal interests should be treated equally unless there is good reason to treat them unequally, but how can we know when interests are truly equal? If I'm on a long, hot hike with my dog and run dangerously low on water, how can I know if my dog's interest in slaking his thirst is greater, less, or equal to mine? Is a chicken's interest in life equal to that of a cow's? Or a mouse's interest in liberty equal to that of a rat's? Significantly, Singer concedes that the lives of some animals have greater inherent value than those of others, and that the lives of persons are generally of greater

2. Peter Singer, "Animals," in Dale Jamieson, ed., *A Companion to Environmental Philosophy* (Malden, MA: Blackwell, 2001), p. 425.

worth than those of nonpersons.[3] This opens the door to some forms of favoritism of human interests over animal interests, but exactly how we should weigh competing interests is a huge problem about which Singer says very little.

Another issue is whether Singer goes far enough in his defense of animal rights and interests. Critics have noted two ways in which Singer's argument may fall short. First, he claims that only sentient creatures have **moral standing** (i.e., have interests or inherent value that "count" from a moral point of view and entitle such creatures to moral consideration and respect). Singer denies that *all* life-forms have inherent value.[4] This seems to imply that there would be nothing wrong with a person secretly wiping out a whole species of complex and beautiful non-sentient organisms simply for sheer fun as long as no harm is done to humans or to other sentient creatures. Second, it is not clear that Singer accords enough protection even to those animals that do have moral standing. He notes, for example, that it might be OK for humans to kill and eat animals (provided no unnecessary suffering is caused)[5] and that some kinds of experiments on animals might be justifiable if the benefits to humans are great enough.[6] These conclusions follow from Singer's commitment to utilitarianism. Though Singer sometimes talks about animal "rights," he really doesn't believe in rights in the traditional sense.[7] His considered view is that sentient animals are entitled to equal moral consideration but not rights.[8] According to Singer, an act is morally right if it maximizes net preference satisfaction, that is, produces the greatest possible balance of satisfied desires over unsatisfied desires.[9] This implies that animals can be killed or exploited by humans whenever this satisfies the greatest number of preferences.[10] For this reason many defenders

3. Singer, *Animal Liberation*, pp. 21–22; and Peter Singer, *Practical Ethics*, 2nd ed. (New York: Cambridge University Press, 1993), p. 107.

4. Peter Singer, "Animals and the Value of Life," in Tom Regan, ed., *Matters of Life and Death* (Philadelphia: Temple University Press, 1980), p. 255n1.

5. Singer, "Animals and the Value of Life," p. 252.

6. Singer, "Animals and the Value of Life," p. 254.

7. Singer, "Animals and the Value of Life," p. 238.

8. Peter Singer, "Animal Liberation or Animal Rights?" *The Monist* 70:1 (January 1987), p. 14; and Peter Singer, "The Parable of the Fox and the Unliberated Animals," *Ethics* 88:2 (January 1978), p. 22. Theorists like Singer who defend animal welfare but reject animal rights are sometimes called "animal welfarists" (as opposed to "animal rightists").

9. Singer, *Practical Ethics*, pp. 13–14.

10. It also implies that *humans* can be killed or exploited by other humans whenever this maximizes net preference-satisfaction. Notoriously, Singer favors the euthanization of severely disabled infants. See Singer, *Practical Ethics*, p. 191.

of animals would argue that Singer's view does not, in fact, accord sufficient protection to animals.

More generally, many critics would claim that the general ethical theory Singer embraces—preference utilitarianism—is not an acceptable moral theory. Preference-utilitarians typically make no distinction between good and bad preferences, counting, for example, the preferences of a sexual harasser as equally weighty as the preferences of his victim. This strikes many critics as wildly unacceptable. Moreover, preference utilitarianism faces serious objections on the issue of justice. Preference utilitarianism is an aggregative theory that looks only at the bottom line, the sum total of satisfied and unsatisfied preferences. As Tom Regan notes, this "seems to imply that good ends justify whatever means are necessary to achieve them, including means that are flagrantly unjust."[11] In some cases, for instance, preference utilitarianism might justify framing an innocent person.[12] For such reasons, many defenders of animals prefer to base their case for animal rights or welfare on nonutilitarian grounds.

A final issue with Singer's argument for animal welfare is whether it has unacceptable anti-environmental consequences. As we noted, Singer believes that only sentient animals have interests, and that nothing is entitled to moral consideration unless it has interests. This view troubles many mainstream environmentalists. It implies, for example, that it would be wrong to hunt invasive goats or rabbits that are threatening endangered plants. It might also imply that it is wrong to reintroduce wolves into Yellowstone National Park and other areas of the American West, because of all the pain and death they are likely to cause elk, deer, rabbits, and other animals on which the wolves will prey. Environmentalists are typically more concerned with the health and integrity of whole ecosystems than they are with the welfare of individual sentient animals within those ecosystems. Singer's view seems to imply that we cannot even favor wild over domesticated animals. Cows and pigs, according to Singer, are to be valued just as highly as wolves and grizzly bears. This strikes many environmentalists as both anthropocentric and ecologically unsound.

Despite these strictures, Singer's contributions are significant and should not be minimized. No one has had a greater impact on the modern animal welfare movement.

---

11. Tom Regan, *Defending Animal Rights* (Urbana, IL: University of Illinois Press, 2001), p. 16.
12. Fred Feldman, *Introductory Ethics* (Englewood Cliffs, NJ: Prentice-Hall, 1978), pp. 57–58.

Regan's Rights View

Singer's utilitarian strategy is one way of defending animal welfare and animal rights. Another is to use a nonutilitarian, duty-based ethics approach. This is the strategy used by Tom Regan (1938–2017) in his influential 1983 book *The Case for Animal Rights*.[13] Regan goes beyond Singer in calling for the total end of animal exploitation, including the use of animals for food, in biomedical research, and in hunting and trapping. Regan bases his defense of animals on what he calls the **rights view**. His central argument can be summarized as follows.

It is widely agreed that all humans have certain basic moral rights, such as the right to life, liberty, bodily security, freedom of thought, and equal treatment under the law. These are rights that everyone is entitled to, regardless of race, color, gender, national origin, or other such morally irrelevant status. What is the basis of such equal basic rights? What is it that *all* humans have in common that grounds such rights? According to Regan, it cannot simply be the fact that we are all members of the human race, because favoring one's own species, simply because it is one's own, is a form of unjustifiable speciesism. The real basis of equal basic rights, Regan argues, is that all humans have equal **inherent value**. By "inherent value" he means "value in its own right." As Immanuel Kant (1724–1804) argued, humans have a kind of "dignity" or "worth" that is intrinsic and not dependent on their quality of life or their usefulness to others. Because all humans have equal inherent value, they should be treated in ways that respect that value and never merely as means to the interests of others or to the collective good of society. Regan calls this basic ethical norm "the respect principle."[14] But what is the basis of humans' equal inherent value? Is it the fact that humans are rational, or autonomous, or capable of making moral choices? No, Regan argues, because some humans (e.g., infants or those suffering from severe dementia) lack all of these attributes. The true basis of equal inherent value, Regan claims, is that all humans are **subjects of a life**. By this he means that humans have beliefs, desires, emotions, memories, intentions, plans, a capacity for pain and enjoyment—in short, a consciously experienced life that matters to them. But humans aren't the only creatures that are subjects of a life in this sense. So, too, are chimps, dolphins, dogs, cats, cows, pigs, and other so-called higher animals. Thus, these animals, like humans, possess equal inherent value and are entitled to respect and basic moral rights. Among these basic rights are the right not to be raised and killed for food as part of a system

13. Tom Regan, *The Case for Animal Rights* (Berkeley, CA: University of California Press, 1983); rev. ed. (2004).

14. Regan, *The Case for Animal Rights*, p. 248.

of commercial animal agriculture, a right not to be hunted or trapped, and a right not to be exploited as a test subject in scientific research.[15]

Regan's argument relies upon a number of widely shared intuitions. The idea that all people possess a kind of dignity or inherent value and are entitled to equal and inalienable basic rights is taught by many religions and is enshrined in historic documents such as the U.S. Declaration of Independence (1776) and the U.N. Universal Declaration of Human Rights (1948). Assuming that Regan is right about these fundamental moral claims, is he also correct in claiming that they apply to some animals as well as to human beings?

Many critics have challenged Regan's claim that being-a-subject-of-a-life is the basis of equal inherent value. A rat may have beliefs, desires, memories, etc., just as a human does, but it cannot do math, make scientific discoveries, write poetry, or make responsible moral choices. Humans, that is, share the psychological traits that make rats subjects of a life, but they also typically possess *additional* capabilities that many would argue give them higher inherent value. Regan also claims that all animals that are subjects of a life are of equal moral worth and must be treated with the same level of respect. This implies that we cannot value dogs over rats, or whales over chickens. If we could somehow determine that a dog and a rat were in an equivalent amount of pain, and we had only one dose of morphine, Regan would presumably say that we should flip a coin to decide which should receive the morphine. Many critics find this strongly counterintuitive.

Regan is on strong ground in claiming that attributes like rationality, autonomy, and moral agency cannot be the basis of humans' equal inherent value, because some humans (e.g., infants and people in an irreversible coma) lack these qualities. But the same can be said of Regan's own proposed criterion of inherent value, being-a-subject-of-a-life. As Regan acknowledges, some humans—for example those in a persistent vegetative coma—are not subjects of a life in his sense.[16] So no one who believes that *all* human lives have equal inherent value can accept Regan's subject-of-a-life criterion.

Is there any attribute that all humans possess, and possess equally, that might serve as a ground for equal inherent value? What about "having an immortal rational soul created in the image of God"? This is an answer that many religious thinkers would favor. Is it satisfactory?

Regan says he is open to the possibility that there are immortal souls but argues that the idea is too controversial to be treated as the basis for the notion

---

15. Tom Regan, "The Case for Animal Rights" in Peter Singer, ed., *In Defense of Animals* (New York: Basil Blackwell, 1985), p. 13.

16. Regan, *Defending Animal Rights*, p. 50.

of equal inherent value or other important moral claims.[17] The idea of immortal souls is certainly controversial, but it is not clear that it should be dismissed for that reason. Most fundamental moral claims—including Regan's own subject-of-a-life criterion—are controversial. Regan's ban on invoking controversial claims would make it impossible to do ethics at a deep level. And if having-an-immortal-soul-created-in-the-divine-image is the only criterion that adequately grounds the notion of equal inherent value, then it must be accepted by those who embrace that view of human equality.

Some critics, including Carl Cohen,[18] have sharply criticized Regan's total ban on animal experimentation. According to the American Medical Association:

> [V]irtually every advance in medical science in the 20th century, from antibiotics and vaccines to antidepressant drugs and organ transplants, has been achieved either directly or indirectly through the use of animals in laboratory experiments. The result of these experiments has been the elimination or control of many infectious diseases—smallpox, poliomyelitis, measles—and the development of numerous life-saving techniques—blood transfusion, burn therapy, open-heart and brain surgery. This has meant a longer, healthier, better life with much less pain and suffering. For many, it has meant life itself.[19]

As Cohen notes, Regan's absolutist position on animal experimentation implies that even if millions of human lives could be saved by experimentation on a single rat, it would be wrong to do so.[20] This is a tough view to swallow, even among defenders of animal rights.

Finally, Regan's view, like Singer's, has consequences that would trouble most environmentalists. He does not permit wild animals like wolves to be favored over domestic animals like cows. His ban on hunting would not permit invasive or overpopulated species of deer or other animals to be culled to prevent ecological damage or mass starvation. His claim that less complex species of animals, such as insects and jellyfish, lack moral standing would be rejected by most

17. Tom Regan, "The Case for Animal Rights," in Carl Cohen and Tom Regan, *The Animal Rights Debate* (Lanham, MD: Rowman & Littlefield, 2001), p. 216.

18. Carl Cohen, "Reply to Tom Regan," in Cohen and Regan, *The Animal Rights Debate*, p. 230.

19. American Medical Association, *Use of Animals in Biomedical Research: The Challenge and Response* (Chicago: American Medical Association, 1992), p. 11; quoted in Hugh LaFollette, "Animal Experimentation in Biomedical Research," in Tom L. Beauchamp and R. G. Frey, eds., *The Oxford Handbook of Animal Ethics* (New York: Oxford University Press, 2011), p. 800.

20. Cohen, "Reply to Tom Regan," p. 230.

environmentalists, who typically believe that *all* life-forms have some intrinsic value. And his focus on the rights of individual animals fits poorly with the kind of "ecological holism" embraced by most environmentalists. As Mark Sagoff notes, "[t]he environmentalist would sacrifice the lives of individual creatures to preserve the authenticity, integrity and complexity of ecological systems."[21] Regan's view would not permit such sacrifices, even at great environmental cost.

3.2 Arguments against Animal Rights

All in all, then, neither Singer's utilitarian argument for animal welfare nor Regan's duty-centered argument for animal rights seems fully convincing. But the idea that *some* animals have *some* moral rights is now widely accepted around the world. Nearly all nations have laws against animal cruelty, and polls indicate growing support for laws that protect animals from needless suffering and exploitation.[22] Are such concerns misplaced? Are there good reasons for thinking that animals have no moral rights at all? Let's look at some common defenses of this claim.

The Contractarian Argument

Some critics of animal rights argue that animals cannot have rights because they cannot make agreements, and therefore cannot be part of any social contract. Here's how conservative commentator Rush Limbaugh puts the point:

> Rights are either God-given or evolve out of the democratic process. Most rights are based on the ability of people to agree on a social contract, the ability to make and keep agreements. Animals cannot possibly reach such an agreement with other creatures. Therefore they cannot be said to have rights.[23]

This argument rests on a flawed moral theory known as **contractarianism**. According to contractarians, moral norms arise from a process of voluntary

---

21. Mark Sagoff, "Animal Liberation and Environmental Ethics: Bad Marriage, Quick Divorce," *Osgoode Hall Law Journal* 22 (1984), pp. 297–307; reprinted in David Schmidtz and Elizabeth Willott, eds., *Environmental Ethics: What Really Matters, What Really Works*, 2nd ed. (New York: Oxford University Press, 2012), p. 63.

22. Rebecca Riffkin, "In U.S., More Say Animals Should Have Same Rights as People," *Gallup*, May 18, 2015. Web. 23 July 2020.

23. Rush Limbaugh, *The Way Things Ought to Be* (New York: Pocket Books, 1992), p. 102.

agreement, a "social contract" in which individuals give up some of their original rights and liberties and agree on a set of binding moral rules for the sake of safety, social order, and the benefits of communal life. Whatever its merits as a political theory of how governments arise and derive their just powers, contractarianism is not an adequate theory of ethics. Contractarianism is a form of moral relativism and is subject to many of the criticisms of relativism that we discussed in Chapter 1. If contractarianism were true, blatantly immoral practices like murder or human sacrifice could become morally acceptable, one and the same act could be ethical in one society and unethical in another, and "non-contractors"—people who didn't or couldn't sign the social contract—would exist in an ethics-free world without either moral rights or duties. Even though infants and severely mentally disabled individuals cannot make and keep agreements, they still have moral rights and deserve to be treated with dignity and respect. Perhaps the same should be said of some or all animals.

The Animal Behavior Argument

In his youth, the American statesman and scientist Benjamin Franklin was a vegetarian, both for ethical reasons and to have more money to buy books. In his *Autobiography*, he describes why he abandoned his vegetarian diet and began to eat meat again:

> [I]n my first voyage from Boston, being becalmed off Block Island, our people set about catching cod, and hauled up a great many. Hitherto I had stuck to my resolution of not eating animal food, and on this occasion I considered . . . the taking every fish as a kind of unprovoked murder, since none of them had or ever could do us any injury that might justify the slaughter. All this seemed very reasonable. But I had formerly been a great lover of fish, and when this came hot out of the frying-pan, it smelt admirably well. I balanced some time between principle and inclination, till I recollected that, when the fish were opened, I saw smaller fish taken out of their stomachs. Then thought I, "If you eat one another, I don't see why we mayn't eat you." So I dined upon cod very heartily, and continued to eat with other people, returning only now and then occasionally to a vegetable diet. So convenient a thing it is to be a *reasonable creature*, since it enables one to find or make a reason for everything one has a mind to do.[24]

24. Benjamin Franklin, *Autobiography* (New York: Walter J. Black, 1941), p. 53.

Franklin's tongue-in-cheek account indicates how he came to accept one of the most common arguments against animal rights: Animals eat other animals, so why shouldn't we?

As Franklin later realized, this is not a sound argument. For one thing, many animals *need* to eat other animals to survive—a lion, for example, cannot eat grass. Also, animals are not moral agents; they act on instinct and cannot discern right from wrong. Humans, by contrast, can recognize the unnecessary suffering they cause by killing and eating animals and act on their compassionate feelings. Finally, why should we think that humans ought to behave like animals? Wild boars wear no clothes, sleep outdoors, and wallow in the mud. Should humans do the same? In general, "Do as animals do" is not good moral advice.

The Line-Drawing Argument

A common rebuttal to claims that animals have rights is: Where do you draw the line? If cows and chickens have rights, why not fish, oysters, mosquitoes, and even microbes? Even if we restrict rights to sentient animals (as Singer does) or subjects of a life (as Regan does), how can we know for certain which animals meet such tests? Since it's impossible to make bright-line distinctions, isn't it better to say that only humans have moral rights?

It is true that virtually any theory of animal rights will face difficult problems of line drawing. But because drawing some lines is hard, it doesn't follow that we should draw no lines at all. It may not always be clear whether it is night or day, but it still makes sense to require cars to use headlights at night. Likewise, it may be hard to know whether, say, a starfish or a jellyfish can feel pain but there is no doubt that a dog or a pig can. This level of precision is enough to get animal-rights arguments off the ground.

The No-Claim Argument

Some critics argue that animals cannot have rights because they can't *claim* rights. The thought here is that rights are essentially valid claims. For example, what does it mean to say that I have a right not to have my car vandalized? It means that if someone trashes my car, I can file charges or claim compensation, and these forms of legal action will be upheld by the proper authorities. But animals cannot make claims. They cannot voice objections or assert entitlements. In short: Rights are valid claims. Animals cannot make claims. So, animals have no rights.

This argument is faulty because, if it were accepted, it would prove too much. Some humans can't make claims either. For example, infants and people with severe mental disabilities can't voice objections, file protests, or assert entitlements. Yet such people still possess rights. They might not be able to personally assert those rights, but others can do so on their behalf. Animals might have rights in a similar way. A dog, for example, might have a right not to be made to suffer without good reason. Even though the dog doesn't know that he has this right and cannot assert such a claim in words, a human surrogate could assert the right for the dog. In fact, this is how many animals are actually treated under the law. Pet owners, for example, typically have a legal duty not to neglect or abuse their pets. The pets can't assert or enforce such duties, but human proxies can do so on the pets' behalf.

The Dominion Argument

One final argument against animal rights was touched on in Chapter 2. We saw there that there are passages in the Bible that report that God has given humans "dominion" over the animals (Gen. 1:28). Other passages state that the "fear" and "dread" of humans "shall be upon every beast of the earth" (Gen. 9:2) and that animals have been "delivered" into human hands for our use (Gen. 9:2). As we saw, such passages have long been cited by Christian thinkers to support claims of human superiority and to deny animal rights. The great medieval thinker Thomas Aquinas, for example, after citing Genesis 9:3 ("I have delivered all flesh to you"), writes: "[A]nimals are ordered to man's use in the natural course of things, according to divine providence. Consequently, man uses them without any injustice, either by killing them or by employing them in any other way."[25]

Is this religion-based argument sound? One issue, of course, is whether the Bible is, in fact, an authoritative text. For non-believers, appeals to the Bible as the revealed word of God will have no force at all. But even if one accepts the authority of the Bible, it isn't clear that it teaches that animals have no rights. For as we saw in Chapter 2, there are passages in the Bible that suggest that God cares for animals and imposes limits on what humans may rightfully do to them. Few biblical scholars today would claim that when the Bible speaks of "dominion" it means "subjugation" or "total domination." For these reasons, it is doubtful that the Bible, read as a whole, provides a good reason for denying all moral rights to animals.

---

25. Saint Thomas Aquinas, *On the Truth of the Catholic Faith: Summa Contra Gentiles*, Book 3, translated by Vernon J. Bourke (Garden City, NY: Hanover House, 1956), p. 119.

3.3 Concluding Thoughts

Where does all this leave us? A strong case can be made that *some* animals have *some* moral rights or moral standing. Science strongly suggests that so-called higher animals, such as dogs and pigs, can suffer and experience emotions such as fear, anger, sadness, anxiety, joy, hope, and grief. As Singer argues, such creatures cannot be lumped with insensate sticks or stones but are living, breathing, sentient creatures that have welfare interests and deserve moral consideration and concern. Along such lines, a strong argument can be made that sentient animals have certain basic rights, including a right not to be killed or made to suffer without adequate justification. The widespread adoption of animal cruelty laws suggests broad support for the idea that some animals, at least, are "morally considerable" and have, or should be accorded, certain fundamental moral rights.

The crucial question is whether we should go beyond such a centrist, commonsense position, and if so, by how much? Should we say that *all* living things (including plants) have inherent value and are thus entitled to at least some level of moral respect? If all living things have inherent value, do they have *equal* inherent value, or can we reasonably make distinctions between higher and lower forms of life? If animals have some degree of inherent value, do humans have more? When animal interests conflict with human interests, how should we resolve such conflicts? These are important questions we will take up in later chapters.

Chapter Summary

1. In this chapter we looked at several common arguments for and against animal rights. Our central question was: Are there good reasons for thinking that at least some nonhuman animals are entitled to be treated with significant moral respect and concern?

2. One influential argument for animal rights (or more precisely for animal welfare) is offered by Peter Singer and is based on utilitarian grounds. Singer argues that because sentient animals can experience pleasures and pains, they have interests (i.e., legitimate welfare-stakes) that humans, ethically, should respect. Morality requires that similar interests be treated similarly; thus, it is wrong to discriminate against animals in

cases where (as is sometimes the case) their interests are equal to our own. Several weaknesses in Singer's argument were noted, including the fact that it seems to have major anti-environmental implications.

3. Another influential argument for animal rights is Tom Regan's duty-centered "rights view." Regan argues that all higher animals are "subjects of a life" in the sense of being sentient beings and having a conscious life and welfare-interests that matter to them. According to Regan, all subjects of a life have inherent value, and in fact equal inherent value, and that is why they possess basic rights, such as a right to life. The main worries with Regan's argument are: (a) whether he is right in thinking that being-a-subject-of-a-life is the basis for equal inherent value, and (b) whether his view, like Singer's, has unacceptable environmental implications.
4. Several common arguments *against* animal rights were also examined. These included: (1) the contractarian argument (animals can't have rights because they can't make agreements), (2) the animal behavior arguments (animals eat other animals, so why can't we?), (3) the line-drawing argument (it's impossible to draw bright lines between nonhuman animals that deserve rights and those that don't, so it's better not to draw lines at all, and deny that *any* animals have rights), (4) the no-claim argument (animals don't have rights because they can't assert rights), and (5) the dominion argument (animals have no rights because God gave humans complete dominion over them and wishes us to use animals for our own benefit). We saw that none of these arguments holds up to scrutiny.
5. The fact that so-called higher animals can suffer and experience emotions such as grief, terror, and fear seems sufficient to ground the claim that some animals have, or should be accorded, certain basic moral rights, such as a right not to be made to suffer needlessly. Whether we should go further and recognize more extensive animal rights is a matter of lively scholarly debate.

Discussion Questions

1. How does Peter Singer argue for animal welfare? Is his argument convincing? What anti-environmental implications does Singer's argument seem to have? Are those implications acceptable?

2. How does Tom Regan argue for animal rights? Is his argument sound? Do you agree with Regan's claim that all higher animals have equal inherent value? Why or why not?
3. What arguments against animal rights did we consider? Are any of those arguments convincing? Are there any good arguments against animal rights that weren't discussed in this chapter?
4. Do animals have certain minimal moral rights, such as a right not to be made to suffer needlessly? If so, do they also have other basic rights, such as a right to life or a right to liberty? What moral rights, if any, do animals possess?

Further Reading

For Peter Singer's defense of animal welfare, see his *Animal Liberation: The Definitive Classic of the Animal Movement*, updated ed. (New York: Harper Perennial Modern Classics, 2009). For Tom Regan's duty-centered defense, see his *The Case for Animal Rights*, rev. ed. (Berkeley, CA: University of California Press, 2004). For a useful collection of diverse perspectives on animal rights, see Cass R. Sunstein and Martha C. Nussbaum, eds., *Animal Rights: Current Debates and New Directions* (New York: Oxford University Press, 2004). For a fascinating and in-depth debate on animal rights, see Carl Cohen and Tom Regan, *The Animal Rights Debate* (Lanham, MD: Rowman & Littlefield, 2001).

Chapter 4

Biocentrism

A central question in environmental ethics is what sorts of things or creatures have moral standing. As noted earlier, something is said to have moral standing (or "moral considerability") if it deserves at least some degree of moral consideration, respect, or concern. Most people would agree that some things in nature have moral standing and some things don't.[1] An ordinary rock, for example, seems to lack moral standing. Rocks can't feel or think or desire or experience pleasure or pain. If you kick them, they can't be injured or harmed. They have no rights that can be violated, no welfare- or well-being interests that can be promoted or diminished.[2] On the other side of the spectrum, nearly everyone would agree that a normal, healthy human baby has moral standing. A baby has wants, needs, interests, and other qualities that make it a being of great moral concern. If babies clearly possess moral standing and rocks clearly lack it, *where do we draw the line*? What sorts of entities belong to the "moral community" and which don't? In short, what are the criteria of moral standing?

For most of Western civilization, it was widely assumed that all and only human beings possess moral standing. Now, as we've seen in previous chapters, many people reject this strongly anthropocentric view. In recent centuries, there has been an *extension*—a widening circle—of moral concern. There's nothing unethical about kicking a fallen pinecone just for fun. Is the same true of a dog? Few would say that it is. Dogs are now widely seen as possessing some level of

---

1. Some people claim that *all* things in nature, both living and nonliving, have moral standing. Those pan-psychists who, like Pocahontas in the Disney film of that name, believe that "every rock and tree and creature has a life, has a spirit, has a name" fall into this category.

2. Although rocks do not deserve any direct moral consideration, they can have various sorts of value (for example, aesthetic, cultural, or economic) that can and should influence the way we treat them. Rocks can also have *indirect* moral standing. It would be morally wrong, for instance, for me to steal a rare and beautiful rock you found in the desert. This is wrong, however, not because of any harm or wrong I do to the rock, but simply because it is your property and you value it. This gives the rock a kind of moral significance that it would otherwise lack. When we ask what sorts of things have moral standing, we are asking whether they have any *direct* moral significance or value.

moral standing. It matters, ethically, what we do to dogs. So a common view today is that humans and some nonhuman animals have moral standing. But again, this raises the critical question: Where should we draw the line?

One fairly popular answer today is: all and only *sentient creatures* have moral standing. Following John Rodman, let's call this view **sentientism.**[3] Leading proponents of sentientism include Peter Singer, Tom Regan, and Joel Feinberg. Feinberg nicely encapsulates the theoretical attraction of sentientism. He writes:

> Without awareness, expectation, belief, desire, aim, and purpose, a being can have no interests; without interests, he cannot be benefited; without the capacity to be a beneficiary, he can have no rights.[4]

Peter Singer makes a similar point:

> The capacity for suffering and enjoying things is a prerequisite for having interests at all, a condition that must be satisfied before we can speak of interests in any meaningful way. It would be nonsense to say that it would not be in the interests of a stone to be kicked along the road by a schoolboy. A stone cannot have interests because it cannot suffer. Nothing that we could do to it could possibly make any difference to its welfare. A mouse, on the other hand, does have an interest in not being tormented, because mice will suffer if they are treated in this way.[5]

In short, Feinberg and Singer claim that only sentient beings have moral standing because only sentient creatures have interests, and only things that have interests matter, morally speaking. Is that argument sound?

The argument, in fact, can be challenged in various ways. Some critics claim that some things possess moral standing even though they have no interests. A human corpse, for example, might deserve respect and proper treatment even though it lacks sentience and can no longer be harmed or benefited. Cutting down a majestic redwood tree to make room for a second hot tub in one's backyard might be morally wrong, not because the tree has interests, but

---

3. John Rodman, "The Liberation of Nature?" *Inquiry* 20 (1977), pp. 83–145. See, generally, Gary Varner, "Sentientism," in Dale Jamieson, ed., *A Companion to Environmental Philosophy* (Malden, MA: Blackwell, 2001), pp. 192–203.

4. Joel Feinberg, *Rights, Justice, and the Bounds of Liberty: Essays in Social Philosophy* (Princeton, NJ: Princeton University Press, 1980), p. 177.

5. Peter Singer, *Practical Ethics*, 2nd ed. (New York: Cambridge University Press, 1993), p. 57.

because it has great intrinsic value that must be properly respected.[6] The claim that only sentient beings have interests can also be challenged. A person in a temporary coma isn't sentient but nonetheless has an interest in being fed and cared for. Non-sentient animals like sponges and oysters have biological needs for food, water, and other things. Don't they have an interest—a welfare stake—in having those biological needs met? Plants, too, seem to have a "good of their own." Sunlight and water are good for flowers, and fire and drought are bad. Since plants can be benefited or harmed, it seems plausible to say that they also have interests.

Convinced by such arguments, many environmentalists today reject sentientism and claim instead that *all* living things, both plants and animals, have moral standing. This is a view commonly known as **biocentrism**. More accurately, biocentrists assert not merely that all life-forms have inherent value (a view also held by many anthropocentrists)[7] but that all living things have *substantive* moral standing.[8] As the label suggests, biocentrists support a life-centered approach to ethics and strongly reject all anthropocentric claims of human superiority or a right to dominate nonhuman nature.

4.1 Albert Schweitzer's Reverence-for-Life Biocentrism

One early and influential advocate of biocentrism was the Nobel Prize–winning doctor and theologian, Albert Schweitzer (1875–1965). In his 1923 book, *The Philosophy of Civilization*, Schweitzer defends an ethical view he called **reverence for life**. Schweitzer believed that all forms of life, plants as well as animals,

6. Alternatively, it might be argued that, even though the redwood tree has no interests, it would be wrong to cut it down because doing so fails to display proper humility, or a proper appreciation of beauty, or some other positive character trait. For an influential argument along these lines, see Thomas E. Hill Jr., "Ideals of Human Excellence and Preserving the Natural Environment," *Environmental Ethics* 5:3 (Fall 1983), pp. 98–110.

7. Aquinas, for example, while holding that whatever God has created is good and represents God's goodness, argues that animals were created for human benefit and may be used and exploited by humans pretty much however we please. See Saint Thomas Aquinas, *Summa Theologica*, Part 1, Q. 47, art. 1 (arguing that all created things are good); and Aquinas, *Summa Contra Gentiles*, Book 3, Chap. 112, para. 12 (arguing that animals were created for human benefit).

8. As we shall see, some biocentrists—namely, biocentric egalitarians—go further and claim that all living things have not only significant moral standing, but also *equal* moral standing.

should not merely be *valued*—they should be *revered* and regarded as *sacred*. As we saw in Chapter 2, many religious and spiritual traditions have embraced similar views. But Schweitzer does not base his view on religion; he appeals instead to logic and intuition.

Schweitzer believed that life is a vast and insoluble riddle. Following the great German philosopher Immanuel Kant (1724–1804), Schweitzer claims that science, philosophy, religion—in fact, all conventional human ways of knowing—are powerless to reveal the true nature and deep structure of reality. All we can know for certain in this confusing world are our own immediate states of consciousness—what we think, how we feel, and how things appear to us at a given moment. When we turn inward and examine our own states of consciousness, the most immediate fact we discover is that *we want to live, and to go on living, and to flourish as deeply and as fully as we can.* This is the first and most important truth about human beings: we are creatures that have a powerful instinctive drive for self-preservation and self-realization that Schweitzer calls the "will-to-live." Moreover, we appear to be surrounded by other life-forms that seem to have an equally powerful drive for survival and self-development. Thus, "[t]he most immediate fact of man's consciousness is the assertion: I am life which wills to live, in the midst of life which wills to live."[9] According to Schweitzer, this is the rock on which all clear thinking about the human condition must rest.

Each of us naturally values and reveres our own life. Can I plausibly claim that my will-to-live is more important than your will-to-live? Or that humans' will-to-live trumps those of other creatures? No, for in all living things we find the same powerful drive for self-preservation and self-fulfillment. It would be arbitrary, therefore, to favor one will-to-live over any other. All life must be revered, and revered equally. Thus, Schweitzer claims that, by a simple "necessity of thought," we arrive at the most basic principle of morality: "It is good to maintain and to encourage life; it is bad to destroy life or to obstruct it."[10]

Schweitzer recognizes, of course, that it is impossible for humans to live without injuring and destroying other life-forms. Every day, we kill countless

---

9. Albert Schweitzer, *Out of My Life and Thought: An Autobiography*, translated by C. T. Campion (New York: Henry Holt and Co., 1933), p. 186.

10. Albert Schweitzer, *The Philosophy of Civilization*, translated by C. T. Campion (Tallahassee, FL: University of Florida Presses, 1981), p. 309. Schweitzer also supports his reverence-for-life ethic by suggesting that it underlies and explains certain common moral attitudes and intuitions (ibid., p. 313). On the face of it, such a claim is implausible (few people "revere" mosquitoes, poison ivy, or tapeworms) and Schweitzer says little to develop or defend the claim.

microorganisms simply by walking down the sidewalk or brushing our teeth. The world is thus a "ghastly drama,"[11] in which life inevitably competes with, and feeds on, other life. The best we can do in such a dog-eat-dog world, Schweitzer argues, is to minimize the harm we do to other living things and assist them when we can. Schweitzer writes:

> A man is truly ethical only when he obeys the compulsion to help all life which he is able to assist, and shrinks from injuring anything that lives. . . . Life as such is sacred to him. He tears no leaf from a tree, plucks no flower, and takes care to crush no insect. . . . If he walks on the road after a shower and sees an earthworm which has strayed on to it, he bethinks himself that it must get dried up in the sun, if it does not return soon enough to ground into which it can burrow, so he lifts it from the deadly stone surface, and puts it on the grass.[12]

So uncompromising is Schweitzer's reverence-for-life ethic that he claims it is *always evil* to destroy or injure life, and anyone who intentionally does so incurs *guilt*.[13] Ethics forbids killing or injuring living things for any reason. If we choose to kill and eat a plant or an animal rather than to starve to death, we have chosen "necessity" over "ethics." When is it permissible to make such a choice? No rules can be given. Each person must decide for himself. As he goes through life and confronts such terrible choices, the ethical person will continually strive to heed less and less the voice of "necessity," and will always be looking for opportunities to make up for the harms he and his fellow-men have committed against other creatures.[14]

Stated compactly, Schweitzer thus argues as follows:

1. I necessarily, and rightly, value and cherish my own will-to-live.
2. All living things have a similar will-to-live.
3. There is no relevant difference between my will-to-live and that of any other living being.
4. Therefore, I should value and cherish all living things—that is, revere life.
5. To "revere life" is to embrace the fundamental moral principle: "It is good to maintain and to encourage life; it is bad to destroy life or to obstruct it."

11. Schweitzer, *Philosophy of Civilization*, p. 312.
12. Schweitzer, *Philosophy of Civilization*, p. 310.
13. Schweitzer, *Philosophy of Civilization*, p. 317.
14. Schweitzer, *Philosophy of Civilization*, pp. 317, 319.

6. Therefore, I should embrace the fundamental moral principle: "It is good to maintain and to encourage life; it is bad to destroy life or to obstruct it."

There is certainly something inspiring about Schweitzer's deeply life-affirming ethics. At the same time, his view has not been widely accepted in environmental ethics. Let's consider several problems with his main argument.

First, some of Schweitzer's premises seem to be false or questionable. Is it true, for example, that *all* living things have a will-to-live? Nothing can have a will unless it has a mind. Does a sunflower, or a microbe, have a mind? Schweitzer believes that they do, claiming, "in and behind all phenomena there is a will-to-live."[15] This suggests that Schweitzer is a kind of pan-psychist who believes that all living and nonliving things are conscious and possess desires, intentions, purposes, and other mental properties. If so, Schweitzer gives no argument for this view. Instead, he seems to base it on a kind of spiritual intuition or mystical insight. If this is his view, then it will be convincing only to those who share his mystical vision.[16]

Another issue: Is it true, as Schweitzer claims, that all wills-to-live are relevantly similar to our own? This claim is crucial to Schweitzer's argument. His claim is that we must revere *all* life because all living things have a will-to-live that is just as imperative as our own. But is that the case? Does a maple tree have an instinct for self-preservation and self-fulfillment? Does it have a conscious desire to avoid death, to go on living, and to flourish in its own distinctively maple-y way? Even if we assume that all living (and nonliving) things have a mind or consciousness, how could we possibly know whether each being's will-to-live is just as strong as our own? Again, Schweitzer's argument seems to rest on a kind of unprovable mystical insight that many would reject.

A similar issue can be raised about Schweitzer's use of the term "reverence." Many environmentalists would agree that all living things are miracles of evolution, have inherent value, and should be treated with some degree of moral respect. But is *reverence* the proper attitude to take toward, say, a flea, a tapeworm, or a dandelion in one's front yard?[17] To regard such organisms as

---

15. Schweitzer, *Philosophy of Civilization*, p. 308.

16. Schweitzer uses the term "ethical mysticism" to describe his reverence-for-life philosophy. Schweitzer, *Philosophy of Civilization*, p. 79. Elsewhere he writes: "All valuable conviction is non-rational and has an emotional character because it cannot be derived from knowledge of the world, but arises out of the thinking experience of our will-to-live, in which we stride out beyond all knowledge of the world." Ibid., pp. 80–81. This embrace of the nonrational is no doubt one reason why his influence on environmental ethicists has been limited.

17. Or "awe," as Schweitzer's German word *Ehrfurcht* is sometimes translated.

sacred, or holy, or as fitting objects of religious awe, seems to make sense only if one considers them to be *divine* or in some way emblems or repositories of the divine. Even then, perhaps, "reverence" may be too strong a term. Can even a mystic sincerely *revere* the lettuce he eats for lunch, the bacteria he brushes from his teeth, or the deer ticks that cause him to contract Lyme disease? Such a response seems neither appropriate nor, for many at least, psychologically possible.

Schweitzer can also be faulted for saying too little about what it means, in practical terms, to revere life. What should we do when human interests conflict with those of other living things? Is it permissible, for example, to build a new public library, given that doing so will destroy the habitats of countless living creatures? Simply saying, "revere life," is unhelpful without at least some general guidelines for resolving conflicts between different life-forms.

Finally, some would claim that Schweitzer's reverence-for-life ethic is unrealistic and overly demanding. Schweitzer's basic moral maxim ("It is good to maintain and to encourage life; it is bad to destroy life or to obstruct it") requires not only that we refrain from unnecessarily harming other life-forms but that we actively do all that we can to aid, promote, and assist life. Schweitzer says, for example, that a truly ethical person would take the time to rescue a worm or an insect that was in distress.[18] This is problematic because, as Buddha reminds us, "life is suffering." Death and pain are ever-present realities in our world. Each day we encounter countless opportunities to "maintain and encourage life." Shall I watch the big game—or volunteer at the animal rescue center? Should I buy some new running shoes—or spend the money on birdseed to feed hungry birds? Schweitzer's claim that an ethical person helps "all life which he is able to assist"[19] requires that each of us become tireless rescuers and benefactors of other living things. Schweitzer accepts this, noting, "the ethics of reverence for life throw upon us a responsibility so unlimited as to be terrifying."[20] "The good conscience," he says, "is an invention of the devil."[21] We must simply do our best, recognizing that this will always be far too little, and accepting our unavoidable guilt. As Schweitzer sees it, such is the tragic human condition in a fallen, pain- and death-filled world.

There are two problems with this response. First, is it sound environmental policy for humans to "play God" and think of themselves as saviors and rescuers of

---

18. Schweitzer, *Philosophy of Civilization*, p. 310.
19. Schweitzer, *Philosophy of Civilization*, p. 310.
20. Schweitzer, *Philosophy of Civilization*, p. 320.
21. Schweitzer, *Philosophy of Civilization*, p. 318.

suffering or dying animals and plants? Isn't it often wiser, from a long-term environmental perspective, to "Let nature take its course"? As Holmes Rolston notes,

> Compassion is not the only consideration in an ethic, and in environmental ethics it plays a different role than in a humanist ethics. Animals live in the wild, where they are still subject to the forces of natural selection, and the integrity of the species is a result of these selective pressures. To intervene artificially in the processes of natural selection is not to do wild animals any benefit at the level of the good of the kind, although it would benefit an individual bison or deer.[22]

Though assistance to distressed organisms is certainly defensible in some contexts, any systematic, large-scale attempts to do so would ultimately wreck the balance of nature, interfere with the course of evolution, and harm species over the long run.

The second problem is whether Schweitzer's reverence-for-life ethic is too rigorous and perfectionistic. Schweitzer was an ordained Lutheran minister who accepted the traditional Protestant doctrine that God demands absolute moral perfection and that it is impossible to perform any act that is strictly "above and beyond the call of duty." Why? Because the Bible teaches that one's duty is always to do what is morally best.[23] This is why Schweitzer claims that ethics condemns *all* intentional killing and injury to life; anyone who does kill or injure abandons "ethics" and chooses instead to act instead from a kind of immoral or non-moral "necessity."[24] This claim is implausible. In ordinary moral discourse it is common to distinguish four categories of moral action: acts that are required or *obligatory* (such as keeping one's promises); those that are *prohibited* (such as lying to make a quick buck); those that are neither required nor forbidden but rather *morally indifferent* (such as choosing to have either an apple or an orange for lunch); and those that are *highly praiseworthy and meritorious but not strictly obligatory* (such as giving half your money to an animal rescue center). Ethicists call deeds of this last sort **supererogatory acts**. Schweitzer denies that there are any such acts. On his view, reverence for life demands that we always do what is absolutely best (namely, intentionally harm *no* life-forms and always do our utmost to assist and maintain life). This seems

---

22. Holmes Rolston III, *A New Environmental Ethics: The Next Millennium for Life on Earth* (New York: Routledge, 2012), p. 73.

23. David Heyd, "Supererogation," *Stanford Encyclopedia of Philosophy*. Web. 17 August 2019. A key text often cited by the Protestant Reformers is Luke 17:10: "So with you: when you have carried out all your orders, you should say: 'We are servants and deserve no credit; we have only done our duty.'"

24. Schweitzer, *Philosophy of Civilization*, p. 317.

MAKING A DIFFERENCE

E. O. Wilson and the Biophilia Hypothesis

For many people today, daily life is lived amidst glass, steel, and concrete, with few connections to nature or to green places. The average American child today spends five to eight hours a day in front of a screen. Is that healthy? There is mounting scientific evidence, in fact, that increasing connections with nature may have a host of psychological and physical benefits.

A leading researcher in this field is the distinguished biologist, Edward O. Wilson (1929–). A two-time winner of the Pulitzer Prize, Wilson believes that human beings have a genetically programmed urge to commune with nature and affiliate with other forms of life. Today, there is a robust scientific research program to explore this so-called **biophilia hypothesis**.

What evidence backs Wilson's biophilia hypothesis? Research suggests that people naturally prefer to look at greenery, flowers, or water rather than at buildings or other artificial constructions. Children seem instinctively drawn to animals. Works of fantasy and science fiction seem to speak to a deep-rooted human hunger to commune with nonhuman life-forms. In addition, psychological studies of children suggest that time spent in nature can reduce stress, boost learning and creativity, alleviate depression, and improve social bonding.

Wilson's biophilia hypothesis has fueled a number of access-to-nature movements. Doctors increasingly recognize the healing powers of nature and prescribe time in the great outdoors to their patients. Cities are working to improve access to green spaces among underserved residents. Nature-based preschools are popping up around the country. Conservation efforts are benefiting from the new evidence of nature's therapeutic effects. And there is a growing international movement to promote universal and equitable access to nature as a fundamental human right.

More than a century ago, environmentalist John Muir proclaimed, "everybody needs beauty as well as bread, places to play in and pray in, where Nature may heal and cheer and give strength to body and soul alike." Wilson's biophilia hypothesis may provide scientific backing for Muir's view.

overly strict and demanding. It implies, for example, that it would be unethical to kill a poisonous snake that was about to bite a child, or to weed one's garden, or to spend money on concert tickets rather than donate the money to an animal rescue center. Even the most fervent nature lovers would find such claims difficult to accept.

4.2 Paul Taylor's Biocentric Egalitarianism

Schweitzer's mystical reverence-for-life form of biocentrism seems implausible, but can biocentrism be defended on other grounds? Philosopher Paul W. Taylor (1924–2015) has offered the most comprehensive and impressive defense of such a theory to date. In his classic 1986 book, *Respect for Nature: A Theory of Environmental Ethics*,[25] Taylor defends an extremely robust form of biocentrism known as **biocentric egalitarianism**. According to biocentric egalitarians, all living things—both plants and animals—have *equal* moral standing and deserve *equal* moral respect and concern. How does Taylor argue for this bold and seemingly extreme biocentric view?

Taylor's argument is complex and proceeds in several steps. First, he argues that all and only living things have moral standing. Why? Because living things have welfare interests and nonliving things, like rocks and piles of sand, do not. Taylor disagrees with Peter Singer that only sentient things have interests. There is a difference, he claims, between *having* an interest and *being aware* of that interest. Suppose, for example, that I need five milligrams of chemical X a day to properly digest my food. Then it's in my interest to get this amount of chemical X even if I'm totally unaware of that fact. According to Taylor, all plants and animals have certain biological needs and interests. A potted plant needs to be watered, and a mouse needs to breathe air. Something has interests if there are things that are good or bad for it, ways that it can be benefited or harmed. All living things have a set of conscious or unconscious welfare interests—what Taylor calls "a good of its own." If something has welfare interests, then it can be benefited or harmed. Whatever can be benefited or harmed matters, morally speaking. Thus, all living things have moral standing. Nonliving things, by contrast, have no welfare interests. There's nothing I can do to a clod

25. Paul W. Taylor, *Respect for Nature: A Theory of Environmental Ethics* (Princeton, NJ: Princeton University Press, 1986). A 25th anniversary edition with a new Foreword by Dale Jamieson was published by Princeton University Press in 2011. My citations are to the original edition.

of dirt that will be good or bad for it. Since nonliving things have no welfare interests, they have no moral standing. Therefore, as biocentrists claim, all and only living things deserve moral concern and consideration.

That is the first step in Taylor's argument. The second is to argue for a general worldview he calls the **biocentric outlook on nature**. This consists of four key claims:

1. Humans are not special or privileged members of earth's community of life; they are parts of that community in the same sense and on the same terms as all other living things.
2. All forms of life on earth, including humans, are interconnected. All the different ecosystems that make up earth's biosphere form a vast web of interdependent relationships.
3. All organisms are **teleological centers of life**. Each, that is, is a unique life-form exhibiting goal-like behavior and pursuing its own good in its own distinctive way.
4. Humans are not inherently superior to other forms of life.[26]

Why should we accept these four components of the biocentric outlook on nature? Taylor's argument for them is complex and multilayered, but essentially he argues that claims 1–3 are highly plausible empirical conclusions in light of modern evolutionary theory and ecology, while 4 is a value judgment that is highly credible given (a) the strong sense of human-nonhuman kinship that emerges from claims 1–3, (b) the lack of good reasons for *affirming* human superiority, and (c) compelling positive reasons for *denying* human superiority.[27] His key reason for denying human superiority and affirming the equal inherent worth of all organisms is that all attempts to show inherent superiority presuppose an indefensible standpoint bias. He writes:

> In what sense are human beings alleged to be superior to other animals? We are indeed different from them in having certain capacities that they lack. But why should these capacities be taken as signs of our superiority to them? From what point of view are they judged to be signs of superiority, and on what grounds? After all, many nonhuman species have capacities that humans lack. There is the flight of the birds, the speed of the cheetah, the power of photosynthesis in the leaves of plants. . . . Why are not these to be taken as signs of their superiority over us?

26. Taylor, *Respect for Nature*, pp. 99—100 (slightly paraphrased).
27. Taylor, *Respect for Nature*, pp. 129—56.

> One answer that comes immediately to mind is that these capacities of animals and plants are not as *valuable* as the human capacities that are claimed to make us superior. Such uniquely human characteristics as rationality, aesthetic creativity, individual autonomy, and free will, it might be held, are more valuable than any of the capacities of animals and plants. Yet we must ask: Valuable to whom and for what reason?
>
> The human characteristics mentioned are all valuable to humans. They are of basic importance to the preservation and enrichment of human civilization. Clearly it is from the human standpoint that they are being judged as desirable and good. Humans are claiming superiority over nonhumans from a strictly human point of view, that is, a point of view in which the good of humans is taken as the standard of judgment. All we need to do is to look at the capacities of animals and plants from the standpoint of *their* good to find a contrary judgment of superiority. The speed of the cheetah, for example, is a sign of its superiority to humans when considered from the standpoint of a cheetah's good. If it were as slow a runner as a human it would not be able to catch its prey. And so for all the other abilities of animals and plants that further their good but are lacking in humans. In each case the judgment of human superiority would be rejected from a nonhuman standpoint.[28]

In short, Taylor argues, any attempt to show that humans are superior in moral standing to other organisms by appealing to humans' allegedly superior capacities will be biased, because it will illegitimately assume that human well-being is the proper standard of judgment. Extending this line of argument, Taylor contends that any attempt to argue for human superiority by appealing to either humans' superior merits or their greater inherent worth will involve a similar kind of question-begging standpoint bias. More generally, since any attempt to demonstrate the inherent superiority of one organism over another would inevitably involve an indefensible standpoint bias, it follows that all organisms must be regarded as having equal moral standing. None deserve more moral respect or consideration than any other. All possess equal inherent moral worth, because the basis of inherent worth is not rationality, or autonomy, or any other allegedly superior capacity or merit, but simply *being a teleological center of life with a good of its own*.[29] This

---

28. Taylor, *Respect for Nature*, p. 130.
29. Taylor, *Respect for Nature*, p. 75.

is something that all (and only) living things possess, and it is a quality they all share equally.

Taylor thus believes that each of the four components of the biocentric outlook on life can be rationally defended. What follows from this? Taylor argues that the biocentric outlook on nature provides a compelling (but not conclusive) ground for adopting an ultimate[30] moral attitude that he calls "respect for nature." This attitude consists of a complex set of character traits, such as dispositions to:

- Regard all wild living things as possessing equal inherent worth
- Desire and promote the good of living beings
- Have life-centered affections, such as feeling angry or unhappy when organisms are unjustifiably harmed or killed
- Perform (or refrain from) acts precisely *because* one sees that they are (or are not) for the good of living beings[31]

The final step in Taylor's theory of environmental ethics is to flesh out, in broad strokes, a system of ethics that comports with an attitude of respect for nature. This system consists of an ordered set of normative principles with three basic elements: (a) standards of good character (i.e., "virtues" that an environmentally responsible person would possess, such as conscientiousness, sensitivity, and self-control); (b) general moral duties that follow from an attitude of respect for nature; and (c) priority rules for resolving conflicts between human interests and those of other living creatures. In general, Taylor's environmental ethic is strongly modeled on theories of human ethics, notably duty-centered theories that focus on concepts like respect, inherent worth, motives, and rights, rather than on utilitarian notions like consequences, preference-satisfaction, or collective welfare. The basic norm of Taylor's theory is that actions are right and character traits are morally good in virtue of their expressing the ultimate moral attitude of respect for nature.[32]

Taylor claims that there are four fundamental ethical duties that flow from an attitude of respect for nature. Simplifying slightly, these are:

- *Nonmaleficence* ("Do no harm to any living thing")

---

30. By "ultimate," Taylor means, "overriding." Like many ethicists, Taylor believes that moral norms always take priority over any competing norms, such as self-interest, aesthetic enjoyment, monetary gain, or scientific progress. Taylor, *Respect for Nature*, pp. 92–93.

31. Taylor, *Respect for Nature*, pp. 80–84.

32. Taylor, *Respect for Nature*, p. 80.

- *Noninterference* ("Don't restrict the freedom of individual organisms" and "Adopt a hands-off policy toward nature and wild ecosystems in general")[33]
- *Fidelity* ("Don't deceive or betray wild animals")
- *Restitutive justice* ("Make restitution to organisms you have wronged")[34]

As Taylor recognizes, such general principles provide only a broad framework for reflecting on one's environmental responsibilities. Since conflicts between human interests and those of other organisms are unavoidable, how should we resolve such conflicts in a way that respects the equal inherent worth of all life-forms? Taylor proposes the following priority principles:

- Self-defense
- Proportionality
- Minimum wrong
- Restitutive justice
- Distributive justice

To apply these principles, we must ask a series of preliminary questions. First: Are the organisms harmful or dangerous to humans? If so, the principle of self-defense applies. This states:

> *Principle of Self-defense*: One can defend oneself (and other humans) against harmful or dangerous organisms that threaten one's life, health, or security.

---

33. Taylor gives a curious reason why humans, as a rule, should not interfere with nature—namely, "that nothing goes wrong in nature." *Respect for Nature*, p. 177. In other words, we should keep our hands off nature because whatever happens in nature is good (or indifferent). In many contexts, this is absurd. If redwood trees were threatened with extinction because of a new blight, should humans stand by and do nothing? If lightning causes a fire that threatens to wipe out the last remaining wild tigers, should we, like the ancient Stoics, say, "Nothing is evil; everything happens for the best"? The notion that nothing that occurs in nature is bad is romantic nonsense and, moreover, wholly at odds with Taylor's rejection of a duty of noninterference in *human affairs*. If, as Taylor claims, humans are part of nature and we should not interfere with nature because whatever happens in nature is good, then it logically follows that humans should not interfere with nature to prevent *human* suffering either.

34. Taylor notes that these are all prima facie or overridable duties, not absolute or exceptionless ones. Each can be overridden by more compelling moral reasons drawn from either human ethics or an ethic of nature. In general, Taylor claims, the four general duties can be ranked in the following order of importance: nonmaleficence, restitutive justice, fidelity, noninterference. *Respect for Nature*, pp. 192–98.

Thus, it would be morally justifiable to shoot a charging grizzly bear, spray insecticide on disease-carrying insects, or apply bacteria-killing antiseptic to a serious wound. However, as with human threats to life, health, or bodily integrity, self-defense must be applied only as a last resort and must be proportional to the seriousness of the threat.[35]

A second threshold question we must ask is: Do the human-nonhuman conflicts involve *basic* or *nonbasic* interests? Basic interests are things that are necessary, or contribute substantially, to an organism's survival or fulfillment of its own good. For humans, basic interests include such primary goods such as life, health, physical security, close personal relationships, rationality, and autonomy. Other organisms will have different basic interests, though all creatures have fundamental interests in remaining alive and in satisfying their biological needs.

A third preliminary question we must ask is: Is the proposed act squarely at odds with an attitude of respect for nature? Does it, for example, treat other organisms as having no inherent worth—as being mere tools for human exploitation or gratification? If so, the principle of proportionality applies. This states:

> *Principle of Proportionality*: Nonbasic human interests must yield to basic nonhuman interests whenever a human act is directly expressive of an exploitative attitude toward nature.

Examples of acts that are "directly expressive" of an exploitative attitude might include: slaughtering elephants for their ivory, killing snow leopards for their furs, using trained tigers in circuses, and hunting or fishing simply for sport.

Sometimes human acts are not directly expressive of an exploitative attitude toward nature but involve an unavoidable conflict between *basic* nonhuman interests and nonbasic (but substantial) human interests. Examples might include clearing land to build a new sports stadium or art museum. In such cases, two moral principles apply:

> *Principle of Minimum Wrong*: Humans may pursue their significant but nonbasic interests, even when this conflicts with the basic interests of other organisms, but only in ways that minimize wrongs to other organisms.

---

35. Taylor dubiously adds that the principle of self-defense does not apply where humans have failed "to make every reasonable effort to avoid situations where nonhuman organisms will be likely to harm us." *Respect for Nature*, p. 268. This implies, absurdly, that it would be wrong to protect a child from an attacking cobra if she had carelessly exposed herself to danger.

Principle of Restitutive Justice: Humans who, in pursuit of either basic or nonbasic human interests, wrong other organisms, must, if possible, make due amends for those wrongs.

The principle of minimum wrong makes sense, Taylor argues, because humans have no duty to return to the Stone Age out of respect for nature and other living things. Basic principles of human ethics permit us to build libraries, universities, parks, housing developments, grocery stores, and other facilities that are necessary to a good quality of human life. At the same time, respect for the equal inherent worth of other organisms requires that when we do allow nonbasic human interests to trump basic nonhuman interests, we do so in a way that minimizes harms to other organisms. An example might be building a new museum on the site of an old factory rather than in an old-growth forest or a pristine mountain meadow. Taylor concedes that determining when it is permissible to allow nonbasic human interests to override basic nonhuman interests can often be a tough call. Ultimately, he says, the choice depends on the value people place on the various interests being furthered.[36] This, of course, will vary from person to person and from society to society. Taylor strongly emphasizes, however, that respect for nature requires a "profound moral reorientation"[37] and seems to impose very challenging constraints on pursuing human goods at the expense of other life-forms.

Finally, we must ask how Taylor thinks we should resolve conflicts between basic human interests and basic nonhuman interests. In such cases, respect for the equal inherent worth of other organisms prohibits any systematic preference for human interests. Some fair or reasonable balance must be worked out. According to Taylor, two moral principles are applicable in cases of basic-vs-basic conflicts. These are:

> *Principle of Distributive Justice*: When basic human interests conflict with basic nonhuman interests, fairness requires that the earth's resources, and more generally all environmental benefits and burdens, be shared equally if possible. In cases where perfect equality is not possible and nonhuman organisms are unfairly wronged, compensation must be paid to the harmed organisms.

36. Taylor, *Respect for Nature*, p. 277.
37. Taylor, *Respect for Nature*, p. 313.

> *Principle of Restitutive Justice*: Humans who, in pursuit of either basic or nonbasic human interests, wrong other organisms, must make amends for those wrongs.[38]

These principles are appropriate, Taylor contends, because basic justice requires that equals be treated equally, and that fair restitution be made whenever a harm is wrongfully inflicted on another person or being that possesses moral standing.

This completes our capsule summary of Taylor's environmental ethics. To recap: Taylor argues for a general theory of environmental ethics he calls "Respect for Nature." Taylor's theory is a form of biocentric egalitarianism, a radically life-centered view of nature that denies human superiority and views all living things as having equal inherent value and equal moral standing. There are three basic components to Taylor's biocentric theory: (1) a set of scientifically informed, life-affirming beliefs about earth's community of life that Taylor calls "the biocentric outlook on nature"; (2) an ultimate (that is, overriding) moral attitude based on the recognition that all living things have equal inherent worth, an attitude that he terms "respect for nature"; and (3) a set of general moral duties, priority principles, and character traits that fit with an attitude of respect for nature. Taylor's theory is certainly impressive, both in its scope and in its argumentative rigor. After more than three decades, it remains the most sophisticated and systematic defense of a thoroughly life-centered environmental ethic ever offered. But is it convincing? Can Taylor's theory—or any form of biocentric egalitarianism—stand up to critical scrutiny?

Critics have raised many objections to Taylor's theory.[39] But the central points of contention are (1) Is Taylor's argument for species equality sound? (2) Does Taylor focus too much on the value of *individual* organisms as opposed to the good of species, ecosystems, and other ecological wholes? and (3) Is Taylor's

38. Since humans unavoidably kill countless organisms just in household cleaning, brushing their teeth, driving to work, and in innumerable other daily activities, any literal application of the principle of restitution would obviously be nonsensical. Consider lawn mowing—a veritable holocaust of flora and fauna. How could I possibly know which plants and animals I killed in mowing and weed-eating my backyard? And how exactly does one "compensate" a heedlessly killed dandelion or ant? Sensibly, Taylor suggests that in many contexts supporting and donating to environmental causes might be the best we can do by way of fair restitution. *Respect for Nature*, p. 188.

39. See, for example, David Schmidtz, "Are All Species Equal?" in David Schmidtz and Elizabeth Willott, eds., *Environmental Ethics: What Really Matters, What Really Works*, 2nd ed. (New York: Oxford University Press, 2012), pp. 114–22; and Joseph R. DesJardins, *Environmental Ethics: An Introduction to Environmental Philosophy*, 5th ed. (Boston, MA: Wadsworth, 2013), pp. 136–45.

ethic overly strict and demanding? That is, does it require levels of human sacrifice that are unrealistic and excessive?

Taylor, as we've seen, totally rejects all claims of human superiority and argues that all living things deserve equal moral respect and concern. He quickly dismisses all religious arguments for human superiority, basing his argument heavily on evolutionary theory, ecology, and other areas of modern science. Anyone who believes that there are sound religious grounds for affirming human superiority will not, of course, be convinced by Taylor's argument for species equality.

Moreover, Taylor's main argument against human superiority is highly questionable. Taylor, as you'll recall, denies that humans have any higher or more valuable traits than other organisms. True, humans characteristically have qualities like rationality, autonomy, and moral agency that other living things lack. But these qualities, Taylor claims, are only "higher" or "more valuable" from a human point of view. From a cheetah's point of view, speed is more valuable than, say, a capacity for moral choice. Thus, Taylor argues, any attempt to defend human superiority (or any claim of species inequality) will be biased and therefore unsound. To give it a name, let's call this Taylor's **standpoint argument**. Is it sound?

There are two serious problems with Taylor's standpoint argument. First, it's false that qualities like intelligence are valuable only from a human point of view. Second, Taylor wrongly assumes that a trait can be valuable only if it is useful to some organism's survival and particular mode of flourishing.

Taylor claims that a trait like high intelligence isn't "really" or "objectively" good—it's only good from a human point of view. That seems clearly false. Many organisms would be benefited if they possessed greater intelligence. A deer, for example, would be far more likely to find food, avoid car collisions, and so forth if it were more intelligent. From a purely biological standpoint, qualities like rationality and critical thinking are generally useful traits to possess. It is true that, from a cheetah's point of view, speed has greater survival value than rationality. A cheetah that traded its speed for intelligence would quickly starve. But that isn't what Taylor needs to show. Taylor needs to demonstrate that no biological traits are objectively "higher" or "better" than any others. And that claim is highly dubious. Intelligence is a useful biological trait, not just for humans but for a wide range of species. What Taylor shows is that intelligence is not a biologically useful trait in all species. That is true, but not what he needs to prove. What Taylor needs to show is that intelligence—and any other allegedly "superior" human trait—has biological value *only* from a human point of view. And such a claim seems clearly false.

Moreover, Taylor's whole argument is based on a false assumption. Taylor assumes that a biological trait has value only if it is useful to some organism. Speed, for example, has value for a cheetah because, and only because, it helps the cheetah survive and live a good cheetah-ish life. This confuses "having value" with "being biologically useful." Being biologically useful is *one* kind of value, but it is not the only kind. A flower, for example, might have aesthetic value even if its beauty confers no biological advantage. The same can be said of certain human qualities that distinguish us from other organisms. Our abilities to write poems or compose music may not be valuable to humans in purely biological terms. They may not increase our odds of surviving, or finding food, or successfully reproducing. But they have other sorts of value. This insight lays bare the crucial flaw in Taylor's standpoint argument. Taylor wrongly assumes that one trait can be superior (= having greater value) to another only if it has greater *biological* value. But biological value is only one kind of value. As David Schmidtz argues, organism A might be inherently superior to organism B because A has all the inherent-value-conferring qualities of B, *plus some additional ones.*[40] This is precisely what traditional defenders of human superiority claim is true of human beings vis-à-vis other organisms on earth. As Aristotle argued, humans share many qualities with plants and animals, but in addition humans can do many other high-value things (like do math, write sonnets, make scientific discoveries, and engage in moral reasoning) that no other organisms can do. True, being able to write sonnets might be of no survival value to a cheetah. But, again, survival value is only one kind of value. Qualities like rationality, moral sensitivity, and aesthetic creativity have value, and not simply from a human standpoint. The fact that a cheetah can neither recognize nor appreciate such values is beside the point. *We* can appreciate them and recognize why they have value beyond simply their biological utility. For these reasons, Taylor's standpoint argument against human superiority is unconvincing.

Critics have also faulted Taylor for being overly focused on the welfare of individual organisms, rather than on the larger good of species or other ecological wholes. As noted in the previous chapter, this is also a common complaint against defenders of animal rights, such as Peter Singer and Tom Regan. Taylor is less vulnerable to this charge than Singer or Regan is because he explicitly concedes that certain groups of organisms have a good of their own and therefore have inherent value.[41] Moreover, as we saw, he also sometimes allows vicarious restitution to be paid to a whole species or biotic community. However,

40. Schmidtz, "Are All Species Equal?" pp. 115–16.
41. Taylor, *Respect for Nature*, p. 18.

Taylor repeatedly argues against any holistic approach to environmental ethics, claiming that we have no reason to care about the good of a species or ecosystem *unless* we see value in individual organisms.[42] In other words, the locus of moral standing in nature lies wholly in individual organisms, and it makes no sense to care about the good of a species or of an ecosystem apart from the good of the individual organisms that make up those collectives.

This argument is unsound. It ignores the familiar fact that, in many contexts, the value of a whole is not simply a function of the value of its parts. The artistic value of a great painting, for example, is not reducible to the value of its individual brushstrokes. Moreover, as Holmes Rolston argues, general patterns or forms ("types") are often rightly more highly valued than individual exemplifications ("tokens") of those patterns or forms.[43] For instance, it would be a much greater evil to destroy all doses of the measles vaccine, and all knowledge of how to make such vaccines, than it would be to destroy a single dose of the vaccine. The same seems be true of ecological wholes such as species and ecosystems. The loss of a single bald eagle is not nearly as great a tragedy as the loss of the entire species of bald eagles would be, for when the species is gone, it can never be replaced, and a long-enduring template of majesty and beauty will have vanished forever from the earth.

Taylor's focus on the good of individual organisms, together with his claim that all living things have equal inherent worth, forces him to oppose many things that most environmentalists today favor. Because environmentalists generally care more about the health and integrity of species and ecosystems than they do about the welfare of individual organisms, they tend to support things like selective culling of overpopulated herds, favoring native organisms over non-native ones, putting out fires that threaten unique or specially valued ecosystems, favoring endangered species over nonendangered ones, privileging animals (in general) over plants, and caring more about "higher" animals (like chimps and whales) than about "lower" animals like mosquitoes and lice.[44] Taylor would reject these claims, or at least struggle to accommodate them within the framework of his species-neutral theory. That is a major problem if, as I would argue, mainstream environmentalists are generally right about what our ecological priorities should be.

The final objection I have to Taylor's theory is that it is unrealistic and overly demanding. Like utilitarianism, it is a "fantasy morality" because it

---

42. Taylor, *Respect for Nature*, pp. 118–19.
43. Rolston, *A New Environmental Ethics*, pp. 134–35.
44. Points made forcefully in Schmidtz, "Are All Species Equal?" pp. 115–17.

imposes unreasonable and excessively stringent ethical demands on people. It does so in two ways.

First, Taylor's theory (like Schweitzer's) moralizes every nook and cranny of our lives. Humans kill and injure countless living things every minute of every day. Each time we eat, breathe, walk, sit down, shower, or brush our teeth "a plaint of guiltless hurt doth pierce the sky,"[45] as innumerable microbes and other organisms are injured or destroyed. Taylor claims that all these organisms are morally considerable and in fact are deserving of the same level of moral concern and consideration that we give to humans. This implies that we must constantly keep our moral radar on high alert. We must continually ask ourselves: What organisms did I harm today? How can I make fair restitution to them? How can I do less harm? Is this a case where I am wrongly favoring my own interests over those of other organisms? Is this a case of basic-vs-basic interests, basic-vs-nonbasic interests, or nonbasic-vs-nonbasic interests? What would a fair distribution of benefits and burdens be in this case? And so on, and so on. Our daily lives would continually be saturated with moral dilemmas, and our moral antennae would be twitching constantly. I contend that it is both unrealistic and inappropriate to require that level of moral attention, reflection, and concern.[46]

Taylor's theory is also a fantasy morality because it requires us to sacrifice too much. Taylor, in fact, is somewhat cagy about what exactly his theory would and would not permit. As Joseph DesJardins notes, it is unclear, for example, whether Taylor's theory would allow me to build a patio in my backyard.[47] On the one hand, digging up the yard would kill lots of plants and small organisms. Basic interests, therefore, would be affected. On the other hand, the reasons why I want to build a patio are so that I can enjoy it with family and friends, drink a few beers out there on fine summer evenings, listen to the birds on crisp spring mornings, and satisfy other nonbasic interests. So, the decision whether to build the patio is a case where *basic* nonhuman interests clash with *nonbasic* human interests. Whipping out my laminated copy of Taylor's normative principles from my shirt pocket, I see that this is a case where the principle of minimum wrong applies. As we noted, this permits humans to favor their own nonbasic interests over the basic interests of other creatures if the nonbasic interests are sufficiently important, so long as minimum harm is done to other organisms and any wrongly harmed organisms are fairly compensated. Given all this, is it ethical to build my patio or not? It's unclear, but I suspect

---

45. Sir Philip Sidney, *Selected Writings* (New York: Routledge, 2002), p. 157.
46. A point made by DesJardins, *Environmental Ethics*, p. 143.
47. DesJardins, *Environmental Ethics*, pp. 143–44.

that Taylor would say No. Building the patio would unjustly privilege relatively trivial human interests over the lives and welfare of innumerable coequal members of earth's biotic community. If this is Taylor's view, then most of us would say that his theory requires humans to sacrifice too much.

To be fair, Taylor would probably be unfazed by this objection. The criticism is based on a widely shared intuition—that people do not have to make significant sacrifices to respect the lives of earthworms, aphids, bacteria, and other common denizens of backyard lawns—that Taylor rejects. In fact, Taylor argues that any appeal to intuitions (that is, unargued assumptions or judgments that just *seem* to a claimant to be true) is illegitimate in ethical reasoning.[48] Argumentatively, this is to Taylor's advantage, because it allows him to dismiss with a wave of his hand a host of seemingly lethal objections to his theory. But the price Taylor pays for excluding all appeals to intuition is too high. Taylor himself, inconsistently, often appeals to intuitions in developing and defending his theory (e.g., in his unargued assertion that it is permissible for humans to protect themselves from dangerous or harmful organisms by destroying them). Moreover, intuitions—though they certainly must be used with caution and due intellectual humility—are indispensable in moral thinking. As C. S. Lewis convincingly argues, without certain bedrock intuitions we have no compass, no measuring stick, to make rationally defensible moral judgments.[49] Without anchoring ethical intuitions, we quickly find ourselves adrift in our moral orientation and can easily fall prey to fantasy or fanaticism. Taylor's own radically life-centered theory seems subject to precisely this charge.

To summarize and conclude: neither Schweitzer's nor Taylor's biocentric theories seem ultimately defensible. Biocentric egalitarianism—the view that all life-forms have equal moral standing and deserve equal moral respect and concern—seems implausible and virtually impossible to practice consistently. At the same time, more moderate forms of biocentrism may be defensible. A plausible case might be made, for example, for the claim that all forms of life have *significant* moral value. A biocentric theory of this type might well be consistent with some versions of **ecocentrism**, or ecological holism, which today is the most popular environmental philosophy. To such theories we now turn.

---

48. Taylor, *Respect for Nature*, pp. 22–24, 269n4.
49. C. S. Lewis, *The Abolition of Man* (New York: Macmillan, 1955), pp. 52–63.

Chapter Summary

1. Biocentrism is the view that all living things have substantive moral value and deserve ethical consideration and respect. Biocentric egalitarianism goes still further and claims that all life-forms have *equal* and substantive moral value.
2. One influential form of biocentrism is Albert Schweitzer's reverence-for-life theory. Schweitzer was a biocentric egalitarian who held that all living things are sacred and should be not merely valued and cherished, but reverenced. He claimed that all living things have a will-to-live that is equal to our own. Because we naturally value and revere our own lives and our own self-fulfillment, we ought, morally, to do the same for all other life-forms. A number of concerns were noted with Schweitzer's argument, including whether it is true that all living things have a will-to-live and whether, even if so, we could ever know that such wills are actually similar to our own.
3. Another important and influential biocentric theory is Paul Taylor's respect-for-nature version. Taylor is a biocentric egalitarian who claims that all living things have significant and equal moral status. Taylor argues that all living things, both plants and animals, have a good of their own and can be harmed or benefited. This implies that all life-forms have moral standing. Moreover, all living things should be treated as having *equal* moral standing because any attempt to show that some life-form is superior to another would be biased and therefore unsound. A number of objections were raised to Taylor's argument. Two major concerns are whether it is true that any claim to superiority or greater moral standing must be unacceptably biased, and whether it is possible to work out a fair and reasonable way to resolve conflicts between human and nonhuman interests that is not unduly demanding in the sacrifices it requires us to make in human ends and values.

Discussion Questions

1. What is biocentrism? How does it differ from biocentric egalitarianism? Do you agree that all living things have substantive moral standing and deserve ethical respect and consideration? If so, why?

2. What is Schweitzer's reverence-for-life biocentric theory? How does he argue for it? Is his argument sound?
3. On what grounds does Taylor argue that all living things have equal moral standing? Is his argument convincing?
4. What is Taylor's biocentric outlook on life? How strong is his argument for that view?
5. Does Taylor succeed in working out a fair and reasonable method for resolving conflicts between human and nonhuman interests? Why or why not?
6. Is Taylor's respect-for-nature ethic realistic? Does it require people to make too many sacrifices to respect the allegedly equal moral status of nonhuman life-forms?

Further Reading

For Schweitzer's reverence-for-life biocentrism, see his *The Philosophy of Civilization*, translated by C. T. Campion (Amherst, NY: Prometheus Books, 1987; originally published in 1923). For Schweitzer's account of how he came to embrace reverence for life, see his *Out of My Life and Thought: An Autobiography*, translated by A. B. Lemke (New York: Henry Holt, 1990). For a sympathetic yet critical analysis of Schweitzer's ethics, see Mike W. Martin, *Albert Schweitzer's Reverence for Life: Ethical Idealism and Self-Realization* (New York: Routledge, 2007). For Taylor's biocentric egalitarian view, see his *Respect for Nature: A Theory of Environmental Ethics* (Princeton, NJ: Princeton University Press, 1986; 25th anniversary edition 2011). For a defense of the biocentric notion that even plants have significant moral status, see Gary E. Varner, *In Nature's Interests? Interests, Animal Rights, and Environmental Ethics* (New York: Oxford University Press, 1998). For a forceful critique of biocentric egalitarianism, see David Schmidtz, "Are All Species Equal?" *Journal of Applied Philosophy* 15 (1998), pp. 57–67.

Chapter 5
Ecocentrism

Over the past few chapters, we've looked at environmental theories that call for an extension of ethical concern. Animal rights supporters like Tom Regan argue that we should extend ethical consideration and moral rights to sentient animals. Biocentrists like Albert Schweitzer and Paul Taylor claim that we should go further—that all living things deserve significant ethical respect and concern. In this chapter, we examine an even more robust form of **ethical extensionism—ecocentrism**. Like biocentrists, ecocentrists reject all strongly human-centered views of nature. Both biocentrists and ecocentrists deny that humans have a right to "conquer" or "subdue" the earth, and that only human beings have intrinsic value or moral standing.[1] But biocentrists and ecocentrists differ on one crucial issue. Whereas biocentrists defend a "life-centered" approach to nature, ecocentrists favor a "nature-centered" or "ecological" or "holistic" view. Put otherwise, whereas biocentrists claim that all and only living things deserve direct moral consideration, ecocentrists contend that some *non*-living things (e.g., mountains, rivers, and wetlands) also have moral standing. More specifically, ecocentrists claim two things: (1) that *both* individual living things *and* certain ecological collectives or wholes (e.g., species, ecosystems, old-growth forests, and earth's biotic community as a whole) have moral standing; and (2) that in environmental decision-making our primary moral concern should be with the health and well-being of ecological wholes, such as species or ecosystems, rather than with that of individual plants and animals. As ethicist Gary Varner notes, ecocentrism is now by far the most widely held view among environmentalists and environmental philosophers.[2] In this chapter, we'll examine the pros and cons of an ecocentric approach to the environment, focusing on the views of Aldo

---

1. As we've seen, some biocentrists (like Paul Taylor) go further and deny that humans are superior in inherent value or moral worth to other living things. Some ecocentrists agree, but most embrace more moderate views.

2. Gary E. Varner, *In Nature's Interests? Interests, Animal Rights, and Environmental Ethics* (New York: Oxford University Press, 1998), p. 11.

Leopold (1887–1948), the founder of ecocentrism and one of America's most influential environmental thinkers. As we'll see, Leopold's holistic environmental theory—what he calls "the land ethic"—is both attractive and problematic. We'll conclude by asking what a more defensible ecocentric ethic, in broad outline, might look like.

5.1 Aldo Leopold's Land Ethic

Leopold was a wildlife expert who studied forestry at Yale, worked for many years for the U.S. Forest Service, wrote the world's first textbook on game management, and taught wildlife management at the University of Wisconsin. His major work, *A Sand County Almanac* (1949), is widely recognized as both a literary and an environmental classic.

Leopold's *Sand County Almanac* is a short, eclectic work consisting of nature essays written over a period of many years. In 1935, Leopold bought a run-down farm near the Wisconsin River in central Wisconsin. There he and his family renovated an old chicken coop ("The Shack") as a weekend and summertime retreat. Many of the essays in Leopold's *Almanac* celebrate the simple joys of nature, the pleasures of ecological literacy, and the joys and frustrations Leopold and his family found in working to restore their farm to ecological health. Much of the book, however, has a prophetic and elegiac tone, as Leopold poignantly chronicles the damage humans have inflicted on once unspoiled forests, prairies, rivers, and other natural landscapes. A recurrent theme in the book is the need to move beyond then-current views of "conservation" and adopt a very different view of humanity's place in the natural order.

As Leopold notes, environmental "conservation" in his day was viewed almost entirely from a human-centered and predominantly economic point of view. "Conservation" was widely seen as a matter of using the earth and its natural resources in a sustainable way in order to promote economic gain and long-term human benefit.[3] National forests, for example, were regarded essentially as "tree farms," to be actively managed so as to produce "the largest possible amount of what crop or service will be most useful, and keep on producing it for generation after generation of men and trees."[4] For similar reasons, pristine wilderness areas were given over to clear-cutting or sheep-grazing; vast

---

3. Donald Worster, *Nature's Economy: A History of Ecological Ideas*, 2nd ed. (Cambridge: Cambridge University Press, 1994), p. 266.

4. Gifford Pinchot, quoted in Worster, *Nature's Economy*, p. 267.

stretches of prairie were plowed up and replaced by agricultural crops; rivers were dammed to provide irrigation, flood control, and hydroelectric power; and wolves, mountain lions, grizzly bears, and other undesirable "varmints" were hunted to the brink of extinction—all in the name of "conservation."

Leopold believed that this strongly human- and economic-centered approach to conservation was inadequate for several reasons. First, modern ecology has shown that natural ecosystems are highly complex and have many interdependent parts. As a result, human attempts to interfere with nature frequently backfire in ways that are difficult to predict. A good example is the systematic extermination of apex predators like grizzly bears and wolves from the American West. As Leopold remarks, if advocates of extermination had been "thinking like a mountain" instead of bowing to pressure from cattle ranchers, they would have realized that predators are crucial for healthy ecosystems because they keep populations of deer and other animals in check, thereby preventing ecological harms like over-browsing, mass starvation of game animals, overpopulation of rodents, soil erosion, and silt-choked rivers.[5]

Second, conservation measures motivated solely by economic gain tend to ignore organisms that have little direct commercial value but may be essential to the ecological health and stability of the land. Beech trees, for example, may be worth less money than many other kinds of trees, but they may play a valuable role in building up soil fertility.[6]

Third, conservation measures tend to be ineffective if property owners are motivated solely by economic concerns. Why, for example, should a profit-minded Wisconsin farmer worry about his topsoil slowly slipping seaward if he can squeeze a few more good crops from his increasingly unproductive land and then move on to a new farm or retire in comfort to the big city? Some might say, "It's the government's job to protect the environment." But Leopold points out that the state can only do so much. Government can encourage sound conservation practices, but its agents cannot be everywhere.[7] Effective conservation must begin with the individual. Farmers and other landowners must recognize that they have *ethical* responsibilities to the land. But how can

5. Aldo Leopold, *A Sand County Almanac with Essays on Conservation from Round River* (New York: Ballantine Books, 1970), pp. 139–40. Leopold also notes that wolves and grizzly bears would ultimately have promoted ecotourism in scenic areas of the West. Thus, it was shortsighted, even on purely economic grounds, to exterminate them (ibid., pp. 144–45).

6. Leopold, *A Sand County Almanac*, p. 249.

7. Implicit, too, in Leopold's discussion is the fact that few Americans in those days favored far-reaching environmental regulations, or "Big Government" in general.

this be achieved when the long-dominant mindset has been that land is simply *property*, a mere *thing* to be exploited for private or public gain?[8]

What is needed, Leopold claims, is a radically new way of thinking and feeling about land and the place of humans in the natural world—a conceptual paradigm-shift that Leopold calls the **land ethic**.

Leopold's land ethic claims that humans have ethical duties not just to their fellow human beings but also to "soils, waters, plants, and animals, or collectively: the land."[9] In other words, humans have moral responsibilities not just to other persons but to *ecological wholes* such as forests, wetlands, grasslands, and entire species of plants and animals as well. According to Leopold, our most basic ethical duty to the environment can be formulated as follows: "A thing is right when it tends to preserve the integrity, stability, and beauty of the biotic community. It is wrong when it tends otherwise."[10] This is what environmental philosopher J. Baird Callicott calls the **summary moral maxim**[11] of Leopold's land ethic. How does Leopold argue for it?

Leopold's core argument for the land ethic rests on two main pillars: science and the history of human moral development. Let's begin with moral development.

As a student of ethics, Leopold recognized that moral attitudes have evolved over time and that they often vary greatly from culture to culture. But Leopold believed that all ethical codes so far developed have rested on one foundational assumption, "that the individual is a member of a community of interdependent parts."[12] Ethics, in other words, is at bottom a matter of cooperation and mutual forbearance for the sake of mutual advantage. Humans are ethical animals because they recognize that society is a web of complex interdependencies and that everyone is better off when people abandon their purely self-regarding pursuits and work together for the common good.

Leopold next draws attention to a deep truth of human moral psychology—that humans naturally care most for family, friends, neighbors, and others with whom they have close bonds or special relationships. Whether for good or ill, humans have evolved in such a way that "[w]e can be ethical only in relation

---

8. Leopold, *A Sand County Almanac*, pp. 244–46.

9. Leopold, *A Sand County Almanac*, p. 239.

10. Leopold, *A Sand County Almanac*, p. 262.

11. J. Baird Callicott, "The Conceptual Foundations of the Land Ethic," in J. Baird Callicott, ed., *Companion to* A Sand County Almanac*: Interpretive and Critical Essays* (Madison, WI: University of Wisconsin Press, 1987), p. 196.

12. Leopold, *A Sand County Almanac*, p. 239.

to something we can see, feel, understand, love, or otherwise have faith in."[13] But therein, Leopold says, lies the ecological rub. Until very recently, prevailing modes of thought have made it impossible to look upon nonhuman life-forms, or the land itself, as part of the human community. But now, Leopold claims, science has made it possible to see and feel our essential kinship with the biosphere itself.

How is this new attitude possible? Leopold cites three major fields of science: Darwin's theory of evolution, geology, and modern ecology.

Leopold believed that "the two great advances" of the nineteenth century "were the Darwinian theory and the development of geology."[14] As environmental historian Roderick Nash points out, Leopold believed that both evolutionary theory and modern geology "helped to tear down the wall Christian thought had so carefully erected between man and other forms of life. The concept of evolution from a common origin over eons of time vividly dramatized man's membership in rather than lordship over the community of living things. On this axiom Leopold built his thought."[15]

Besides Darwinism and geology, the third big intellectual catalyst making possible a shift to a radically new environmental vision is modern ecology—specifically, the so-called New Ecology that emerged in the 1930s and 1940s.[16] The central premise of the New Ecology was the intricate complexity and web-like interdependence of natural ecosystems. In nature, as Barry Commoner was later famously to say, "everything is connected to everything else."[17] A vivid illustration of this is what Leopold calls "the land pyramid."[18] Ecology has shown that all life on earth depends on food chains, with plants absorbing energy from the sun, insects and smaller animals feeding on plants, larger carnivores feeding on insects and smaller animals, and so on, up the food chain to apex predators such as lions and grizzly bears. In scientific terms, all of this constitutes a kind of pyramid-shaped, solar-powered "energy circuit" in which plants outnumber smaller animals, and smaller animals outnumber larger animals, and in which energy is continually recycled through

13. Leopold, *A Sand County Almanac*, p. 251.

14. Aldo Leopold, "Wilderness" (unpublished manuscript, c. Dec. 1935), in Curt Meine, ed., *Aldo Leopold:* A Sand County Almanac *and Other Writings on Ecology and Conservation* (New York: Library of America, 2013), p. 375.

15. Roderick Frazier Nash, *Wilderness and the American Mind*, 4th ed. (New Haven, CT: Yale University Press, 2001), pp. 193–94.

16. Donald Worster, *Nature's Economy*, pp. 293–315.

17. Barry Commoner, *The Closing Circle: Nature, Man, and Technology* (New York: Alfred A. Knopf, 1971), p. 33.

18. Aldo Leopold, *A Sand County Almanac*, p. 251.

nature and eventually returned to the soil and to the oceans when plants and animals die and decompose. From an ecological standpoint, Leopold points out, there is nothing special about humans in this energy circuit of soils, waters, plants, and animals. Like all life on earth, we depend on the sun and on food chains that descend from the largest predators down to the tiniest plants and microorganisms. Thus, from both an evolutionary and an ecological standpoint, humans do not stand above and apart from nature. Rather, we are coevolved earth-dwellers who should view our role not as conquerors of the biotic community but in Leopold's famous phrase as "plain citizen[s] and member[s] of it."[19]

In sum, Leopold argues, modern science, aided by the natural human tendency to care most for kin and community, has, for the first time in Western civilization, made a new ecological ethic of respect for the land a real possibility. But there is also an additional factor at play, Leopold believes. Since ancient times history has revealed an expanding circle of ethical concern. In a memorable passage, Leopold recounts the scene from Homer's *Odyssey* in which "God-like Odysseus" returns from Troy and hangs twelve allegedly misbehaving slave-girls all on one rope.[20] Leopold notes that in Homer's time (roughly 750 BCE), this raised no ethical issue, since slaves were then considered as simply property, deserving neither rights nor ethical regard. Since Homer's time, there has been a dramatic expansion of what ethicists term "moral standing" (or "moral considerability"). Civil and political rights have gradually been extended to women, the working class, and to an ever-widening circle of racial, ethnic, and cultural minorities. As we have seen, some environmentalists argue that moral and legal rights should be extended to sentient animals. Has this process of ethical extension come to a stop or will it continue? Leopold believes that it can and will go further. In Leopold's view extending ethical consideration to "the land"—that is, to entire ecological communities of soils, waters, plants, and animals—is now a real "evolutionary possibility."[21]

But there is more. A new land ethic, Leopold argues, is not merely possible; it is urgently needed. Given the alarming pace of soil erosion, habitat destruction, wildlife loss, resource depletion, pollution, and other forms of environmental degradation, a radically new way of regarding the land is now "an ecological necessity."[22] We can no longer afford to think of the land simply as a

19. Aldo Leopold, *A Sand County Almanac*, p. 240.
20. Aldo Leopold, *A Sand County Almanac*, p. 237.
21. Aldo Leopold, *A Sand County Almanac*, p. 239.
22. Aldo Leopold, *A Sand County Almanac*, p. 239.

resource to be exploited and plundered. We must begin to think of it as a community, a common home to be sustainably shared with all of our co-evolved, co-dependent fellow creatures.[23]

In a nutshell, Leopold's core argument for the land ethic runs as follows:

1. Humans are now causing great and often irreversible damage to the environment. (Evidence: current trends on habitat and wildlife loss, soil erosion, resource depletion, pollution, and so forth.)
2. Thus, it is an ecological necessity to abandon traditional ways of viewing nature and adopt a radically new ecological ethic—an ecocentric "land ethic" that recognizes the complexity and intrinsic value of ecosystems; regards humans as plain citizens and members of earth's biotic community; and recognizes a basic ethical duty to preserve the integrity, stability, and beauty of the land.
3. By nature, humans tend to care most for family, friends, and community—in short, those they perceive as kith and kin.
4. Recent advances in science (notably ecology, geology, and evolutionary theory) and in human ethical development now make it possible for us to regard the land itself—and all its biota—as kith and kin.
5. Thus, it is now a real psychological possibility to adopt a new ecocentric land ethic.
6. Therefore, we both can and should adopt a new ecocentric land ethic.

This, in barest outline, is Leopold's argument for the land ethic. Is it convincing?

Today, Leopold's land ethic is widely embraced by environmentalists. Like Leopold, most mainstream environmentalists now strongly reject human-centered views of nature, hold that all living things have intrinsic value, and believe that humans need to achieve a healthy and sustainable balance with nature. Most environmentalists today also adopt a broadly ecocentric view of nature. That is, they believe that our primary environmental concern should not be with the welfare of individual plants or animals but with the good of

23. My summary of Leopold's argument for the land ethic is greatly indebted to Callicott's "The Conceptual Foundations of the Land Ethic." Callicott argues that Leopold's argument also invokes "Copernican astronomy"—specifically the idea that humans and other terrestrial organisms are cosmologically insignificant fellow travelers on Spaceship Earth (and thus, so to speak, "all in this boat together"). This may be a good reason for feeling a sense of solidarity with nature, but it is not a reason offered in *A Sand County Almanac*. Here, as elsewhere, some aspects of Callicott's interpretation are highly speculative with little clear basis in Leopold's writings.

ecological wholes, such as species and ecosystems. Like Leopold, most contemporary environmentalists support things like:

- Prioritizing wild species (e.g., wolves) over domestic ones (like cows)
- Favoring endangered species over thriving ones
- Giving precedence to native species over exotic or invasive ones
- Prioritizing ecologically valuable species (e.g., honeybees) over those that are less useful or even damaging to the environment (e.g., tree-killing insects or fungi)
- Giving precedence to animals (especially putatively "higher" or more complex animals, such as elephants and jaguars) over plants
- Permitting selective culling and sport hunting to keep overpopulated or invasive animals in check
- Rejecting radical environmental theories (e.g., views that call for a rapid re-wilding of vast portions of the globe) and working to achieve a fair and sustainable balance between human and nonhuman interests

In short, unlike many of the environmental theories we have examined so far, much of the spirit of Leopold's land ethic has actually been absorbed into mainstream environmental thought.[24]

That said, we must still ask whether Leopold's central claims are true and whether his arguments for those claims are sound. Let's begin by considering some common objections to the land ethic.

One frequent complaint is that Leopold isn't very clear on exactly what the land ethic amounts to, or on why we should embrace it.

Consider the linchpin of his land ethic, his so-called summary moral maxim: "A thing is right when it tends to preserve the integrity, stability, and beauty of the biotic community. It is wrong when it tends otherwise." How exactly should that be understood? Here are a few clarifying questions we might wish to ask:

- What is the *scope* of Leopold's summary moral maxim? Is it the basic moral principle of *all* ethics, or does it apply only to issues of

24. See, for example, Elliott Sober, "Philosophical Problems for Environmentalism," in Bryan Norton, ed., *The Preservation of Species* (Princeton, NJ: Princeton University Press, 1986); reprinted in Schmidtz and Willott, eds., *Environmental Ethics*, p. 133 (noting, "what is special about environmentalism is that it values the preservation of species, communities, or ecosystems, rather than the individual organisms of which they are composed"). Or as Callicott more colorfully expresses the point: "Leopold has street cred in the environmental movement 'hood like nobody else, not Thoreau, not Muir, not Pinchot." J. Baird Callicott, *Thinking Like a Planet: The Land Ethic and the Earth Ethic* (New York: Oxford University Press, 2013), p. 13.

land management or environmental decision-making?[25] (An act of bank robbery, for example, may have no real impact on any "biotic community" of soils, waters, etc. Is it thus neither right nor wrong?) In other words, is the principle restricted to environmental matters, or, since virtually all human activities affect "biotic communities," is it intended to have a broader application?

- What is the *force* of Leopold's summary moral maxim? Is it intended as "an ethical absolute" with no allowable exceptions? Or does it impose only "prima facie" or overridable moral obligations? If the latter, what sorts of exceptions are permitted? Would it, for example, allow for construction of a new soccer stadium in a drained wetland, or a swimming pool in a public park?
- Leopold claims that the land ethic is an extension of earlier moral values. (For example, it extends moral status from human beings to soils, waters, plants, animals, and so on.) If so, what is the relationship between the outer and inner spheres of this expanding circle of moral concern? Do the inner and outer circles perfectly harmonize in their ethical implications, or do they sometimes conflict? If the latter, how should such conflicts be resolved?
- What exactly is a "biotic community"? Is a seasonal pond, or the underside of a fallen log, or a compost heap a biotic community? What about something much larger, like the Arctic tundra or the Indian Ocean? If biotic communities are to be valued and preserved, as Leopold urges, it must be possible to pick them out from other parts of the natural world. But how can this be done?
- What does it mean to preserve, or fail to preserve, the "integrity" of a biotic community? What is it for, say, a beaver pond or a sand dune to possess "integrity"? If I cut down some maple saplings in my backyard and replace them with flowering fruit trees, have I wrongly damaged the "integrity" of a biotic community?
- However they are defined, the values of "integrity," "stability," and "beauty" would seem to be different and potentially conflicting standards. If I plant azaleas in my backyard, might this increase the beauty of my yard but somewhat decrease its ecological stability and/

---

25. Don Marietta labels the former view, that duties to the biosphere override all other duties, "extreme deontic holism." Don Marietta Jr., "Environmental Holism and Individuals," *Environmental Ethics* 10:3 (Fall 1988), p. 254. For an early attempt to flesh out the meaning of Leopold's summary moral maxim, see James D. Heffernan, "The Land Ethic: A Critical Appraisal," *Environmental Ethics* 4 (1982), pp. 235–47.

> or integrity? How should we resolve conflicts between the potentially competing ecological values of integrity, stability, and beauty?

As these questions suggest, it is far from clear how to interpret Leopold's summary moral maxim, or what its practical implications are for environmental decision-making.

Questions can also be raised about the logic of Leopold's argument. As we saw, Leopold clearly believes that science—especially ecology, geology, and evolutionary theory—in some sense "supports" the land ethic. But the land ethic is an *ethic*—a set of normative values and commitments. Science deals with observable facts, not values. As the philosopher David Hume (1711–1776) rightly pointed out, no "ought" (i.e., value statement) can logically be derived from an "is" (i.e., purely factual statement). So what exactly is the logical or evidentiary connection between the scientific facts Leopold cites and the ethical conclusions he draws from them? Since science by itself can't support those ethical inferences, what else is Leopold assuming that would provide the necessary support? And are those assumptions correct? These are important questions that Leopold leaves unanswered.[26]

So one common—and valid—complaint is that Leopold should have been clearer on what the exactly the land ethic is and why we should accept it. What other objections have critics raised?

One of the most striking—and novel—things about Leopold's land ethic is its "holism"—the claim that we have ethical duties to species, ecosystems, and other ecological collectives, and that our main ethical concern should be with the well-being of those wholes, not with the welfare of individual plants or animals. Leopold's holistic value system raises two key questions: (1) Is it true that we owe ethical duties to ecological wholes? and (2) Does Leopold's holism amount to a kind of "environmental fascism"—an indefensible subordination of the good of the individual to the good of the whole?[27]

The notion that we owe ethical duties to nonliving wholes such as mountains and deserts seems puzzling, because such things presumably lack minds

---

26. Valiant attempts have been made by Leopold's leading interpreter and defender, J. Baird Callicott, to address such questions—to fill in the gaps, as it were, in Leopold's land ethic (with mixed success, in my opinion). In fairness, Leopold was not a professional philosopher and was writing for general readers. Still, he can be dinged for not explaining his view more fully and clearly. The interpretive uncertainties are quite serious.

27. Tom Regan, *The Case for Animal Rights* (Berkeley, CA: University of California Press, 1983), pp. 361–62; and William Aiken, "Ethical Issues in Agriculture," in Tom Regan, ed., *Earthbound: Introductory Essays in Environmental Ethics* (Prospect Heights, IL: Waveland Press, 1984), pp. 269–70.

and cannot feel pain or care about what happens to them. On the other hand, we all talk quite naturally about doing things "for the good of the team" or doing what is "in the best interests of the country," where we assume that there is some aggregate or collective entity that has interests that are not fully reducible to the interests of the individuals that compose them. Whether collective entities, such as old-growth forests, prairies, or species, have direct, nonreducible moral standing is an ongoing debate in environmental ethics.[28]

Perhaps a more serious worry is the charge of eco-fascism. Does the land ethic place too much value on the good of collectives and too little on the welfare of individual organisms? Significantly, Leopold's summary moral maxim says nothing about individuals but speaks only of the good of "the biotic community." As Elliott Sober notes, this is troubling, because it might justify the infliction of enormous suffering on individual animals (or, in fact, on individual humans) if the good of a species or of an ecosystem required it.[29] The fascist idea that the individual is nothing and the good of the state (or "the People" or "the race") is everything, is rightly condemned in human affairs. Why should we accept it with respect to nature? Put otherwise: Why should we accept an *environmental* ethic that seems to be radically opposed to any defensible *human* ethic?

Some defenders of Leopold have denied that the land ethic implies any form of fascism.[30] But the notion that individual plants and animals matter less, ethically and ecologically, than species and other ecological wholes is inherent in the very concept of ecological holism. It remains a fair question, therefore, whether Leopold's land ethic strikes the right balance between ecological wholes and the individual organisms that compose them.

A final criticism of Leopold's land ethic should also be noted. Leopold's summary moral maxim says that an act is right if it tends to *preserve* the integrity, stability, and beauty of biotic communities. This notion of preservation

---

28. For an affirmative view, see Lawrence E. Johnson, *A Morally Deep World: An Essay on Moral Significance and Environmental Ethics* (New York: Cambridge University Press, 1991), pp. 202–8; and Holmes Rolston III, "Ethics on the Home Planet," in Anthony Weston, ed., *An Invitation to Environmental Philosophy* (New York: Oxford University Press, 1999), pp. 129–30. For a negative view, see Peter Singer, *Practical Ethics*, 2nd ed. (New York: Cambridge University Press, 1993), pp. 280–84; and Harley Cahen, "Against the Moral Considerability of Ecosystems," *Environmental Ethics* 10:3 (1988), pp. 195–216.

29. Sober, "Philosophical Problems for Environmentalism," p. 133.

30. See, for example, Callicott, *Thinking Like a Planet*, pp. 65–66. Callicott plausibly argues that Leopold held that human ethics—including the rejection of fascism in our treatment of our fellow human beings—often trumps ecological duties to nonhuman biotic communities.

raises several issues. Modern ecology has shown that randomness, disturbance, and change, not stability, is the norm in most natural ecosystems.[31] Ice Ages come and go, oceans rise and fall, grasslands burn and recover, deserts expand and contract, superstorms devastate the land, plagues of insects kill off whole forests, non-native species colonize new territories, and there is constant and often random successional change of plants and animals in existing ecosystems.[32] If humans have an ethical obligation to preserve the "integrity" and "stability" of natural ecosystems, won't frequent and massive human interventions be needed to restore ecosystems that have been seriously disrupted by human or natural causes? To do so would be contrary to the general hands-off approach to nature most environmentalists favor. (It would also, of course, be hugely expensive.) In addition, if it is wrong for humans to disrupt the integrity or stability of natural ecosystems, severe restrictions would need to be imposed on human economic activity. No longer could we dam rivers for hydroelectric power, clear land to build housing developments, or plow up prairies to grow food. This apparently opens Leopold to the charge of *misanthropy*—of unfairly favoring nonhuman interests over human ones and imposing excessive constraints on human progress.[33] Finally, aren't there times when it is legitimate to seek to *improve* nature rather than simply to preserve it? Damming a river may transform an arid desert into a verdant oasis. Planting flowering trees or clover may increase an area's beauty, fertility, and biodiversity. Preservationist bromides such as "Nature knows best" are, at most, only half-truths. No plausible case can be made that nature's status quo must *always* be preserved, "even though the heavens fall."

In sum, there are some serious gaps and other problems with Leopold's ecocentric land ethic. Still, as we noted earlier, something along the lines of Leopold's land ethic is widely accepted in mainstream environmental thought today. This raises an important question: Can Leopold's land ethic—with all

---

31. Donald Worster, "The Ecology of Order and Chaos," *The Environmental History Review* 14:1–2 (March 1990), pp. 1–18.

32. Worster, *Nature's Economy*, pp. 388–433; and Callicott, *Thinking Like a Planet*, p. 96.

33. See, for example, Louis P. Pojman, *Global Environmental Ethics* (Mountain View, CA: Mayfield Publishing Co., 1999), pp. 161–62; and Aiken, "Ethical Issues in Agriculture," pp. 169–70. For the record, I believe the charge of misanthropy is unfair as applied to Leopold. As Mark Bryant Budolfson convincingly argues, it is a mistake to see Leopold as a radical environmentalist. Rather, he is best viewed as a moderate anthropocentrist. See Mark Bryant Budolfson, "Why the Standard Interpretation of Aldo Leopold's Land Ethic is Mistaken," *Environmental Ethics* 36:4 (Winter 2014), pp. 443–53.

its acknowledged problems—be revised (clarified, expanded, shored up) and (if necessary) modified to render it a fully defensible environmental ethic? I suggest that it can. Here and in the following chapter, let's briefly consider what a more satisfactory version of Leopoldian ecological holism might look like.

5.2 A Modified Land Ethic

Leopold's land ethic has many virtues. As noted earlier, ecocentrism is currently the most popular environmental ethic. Many contemporary environmentalists like the land ethic because it rejects extreme forms of anthropocentrism, strikes what many see as a reasonable balance between human and nonhuman interests, teaches that nature should be valued and respected for both human-centered and intrinsic reasons, avoids environmental extremism, and favors policies that most environmentalists support (such as species and wilderness preservation, a focus on the good of ecological wholes, and favoring native species over invasive ones). These are attractions that any refurbished land ethic should preserve. But as we saw, there are serious issues with the way Leopold formulates and defends the land ethic. How should those problems be addressed?

We saw that many of the problems we pointed to relate to Leopold's summary moral maxim ("A thing is right when it tends to preserve the integrity, stability, and beauty of the biotic community. It is wrong when it tends otherwise"). As we noted, it is far from clear how this should be understood. Is it the fundamental principle of all ethics or only of ecological ethics? Is it a moral absolute or does it have exceptions? What is a "biotic community"? How can we individuate such communities? What does it mean to speak of the "integrity," "stability," and "beauty" of a species or of an ecosystem? How can these criteria be measured? How should we resolve conflicts between these criteria? Finally, how seriously should we take Leopold's admonition to "preserve" the beauty, stability, and integrity of species and ecosystems? Does this require constant and large-scale human interventions in nature? Does it permit us to alter or destroy natural areas for important human purposes? And are there times when it's OK for humans to seek to heal, beautify, or otherwise improve nature rather than to leave it untouched?

Because of these various difficulties, I believe that Leopold's summary moral maxim must be either substantially revised or replaced with something else. The best solution, I think, would be to abandon the search for any single "summary" ecological "maxim" and instead adopt a plurality of general principles. This is the approach taken in virtually all leading environmental documents, including the much-admired "Earth Charter" (2000).[34] "The Earth Charter" invokes sixteen general principles, including these:

- Respect Earth and life in all its diversity.
- Care for the community of life with understanding, compassion, and love.
- Secure Earth's bounty and beauty for present and future generations.
- Protect and restore the integrity of Earth's ecological systems, with special concern for biological diversity and the natural processes that sustain life.

Similar principles could be formulated to reflect the basic vision and values of Leopold's land ethic. To reflect the broadest environmental consensus, it should be made clear that the principles are not exceptionless, that they should interpreted reasonably and not with wooden literalness, and that their intent is to promote widely shared environmental values and to achieve a just and sustainable balance between human interests and those of other members of Earth's community of life.

In addition to problems with Leopold's summary moral maxim, a second major concern with the land ethic is how Leopold tries to support it. As we saw, Leopold rests his land ethic largely on science. His claim, baldly stated, is that that we can and should adopt the land ethic, because we naturally care most about kin and community, and science has shown that "the land" is, in relevant ways, our kin and community. This argument fails for two major reasons. First, as noted earlier, scientific facts alone cannot support any normative conclusion; no "ought" can be logically inferred from an "is." Leopold needs a value premise to make his argument work, but it isn't clear what premise would do the trick or why we should accept it.[35] Second, it is false to say that science has shown that soils, waters, and so forth are "kin" or "community" in any sense robust enough to support Leopold's argument. Science has indeed shown that humans share a

---

34. "The Earth Charter." Web. 29 December 2019. The Earth Charter is a non-binding declaration of fundamental sustainable-development principles developed through an international consultative process as a nongovernmental initiative in the aftermath of the Earth Summit (1992). The full text, in various translations, can be found on the Earth Charter website.

35. One obvious candidate would be: "It is right to care about kin and community." But this is at best only generally true, and what is its logical grounding? Given Leopold's appeal to evolution, he might argue that any moral instinct or response instilled in us by evolution is not only natural, but also right. However, that would commit him to a form of evolutionary ethics, with all its well-known problems. So, it remains unclear how or whether the lacuna in Leopold's argument can be repaired.

common evolutionary ancestry and genetic endowment with all other living things on earth. (Humans share more than 40% of their DNA with bananas.)[36] But soils and waters are not living things. In what sense, then, are they "kin"? Science has also shown that, broadly speaking, "everything is connected to everything else" in the sense that all terrestrial life is part of an interlocking web of mutual dependencies. In this thin sense, humans are indeed part of a "community" of life-forms and their environments. But this sense of "community" isn't strong enough to support Leopold's argument. For one thing, nature is "red in tooth and claw"—a *Hunger Games*–like battlefield of incessant competition, strife, killing, and eating.[37] The kind of "community" that exists between a lion and a gazelle is surely a rather attenuated one. Moreover, it may be true that evolution has wired humans so that we naturally feel moral connections for *other humans* that are enmeshed with us in webs of mutual dependency. But the same cannot be said for other kinds of "communities" of dependency. A villager may be dependent on his ox, and vice versa, and yet the villager may be indifferent to the ox's suffering. A farmer may depend on his bean crop, and the bean crop may depend on the farmer, and yet the farmer may feel no moral concern for the bean plants at all. For these reasons, Leopold's attempt to ground his land ethic on science must be accounted a failure.

A better argument for the land ethic must be multipronged. It can and should include relevant scientific facts from ecology, evolutionary psychology, evolutionary theory, biochemistry, astrophysics, and other areas of science. But these should be part of a larger normative argument aimed at supporting a scientifically informed holistic ethic. The central thrust of the land ethic is that we should care for and protect the land *both for our sakes and for the sake of the land itself*. A convincing case for the land ethic, therefore, must include both anthropocentric and nonanthropocentric reasons for protecting the environment. What must be shown, in short, is that *both* humans and nature matter, that our origins and destinies are closely intertwined, and that we must find a *modus vivendi* in which both may survive and thrive. A fuller account of what a defensible ecocentric ethic might look like will be presented in Chapter 7.

---

36. Elanor Cummins, "Bananas: Your Cousin, Maybe?" *Popular Science*, August 14, 2018. Web. 14 June 2020.

37. A point noted in Rolston, "Ethics on the Home Planet," p. 128. Rolston, however, argues that ecosystems are "communities" in an extended but nonetheless legitimate sense.

Chapter Summary

1. Ecocentrism (or "ecological holism") is the view that ecosystems, species, and other ecological wholes have moral standing, and that in environmental decision-making our primary concern should be with the health and well-being of such wholes, rather than with that of individual plants and animals.
2. The founder and most influential proponent of ecocentrism is Aldo Leopold, author of the environmental classic *A Sand County Almanac* (1949). Leopold defends a version of ecocentrism he calls "the land ethic." The land ethic claims that we have ethical responsibilities to the land itself; that humans should not see themselves as privileged citizens of earth's community of life but rather as plain members; and that our primary ecological duty is to preserve the integrity, stability, and beauty of the biotic community. Most environmentalists today are ecocentrists.
3. Leopold argues that we should extend ethical concern and respect to rocks, soils, waters, plants, animals—in short, to "the land" itself. He contends that modern science—especially modern ecology—has shown that life on earth is a "community," an interdependent whole. Given the pace of environmental destruction and the holistic view of nature that modern science has revealed, it is now both an ecological necessity and a real psychological possibility to extend ethical concern to the land itself.
4. Criticisms of Leopold's land ethic center on whether he succeeds in showing that nonliving ecological wholes, such as entire ecosystems, deserve direct ethical concern, whether his mostly science-based argument for the land ethic succeeds, and whether his proposed "summary moral maxim"—to preserve the integrity, stability, and beauty of the biotic community—is an adequate basis for an ecocentric ethic.

Discussion Questions

1. What is ecocentrism? How does it differ from biocentrism? Which view is more defensible?
2. What is Aldo Leopold's land ethic? How does he argue for it? Do you find his argument convincing?

3. What is Leopold's summary moral maxim? How should it be interpreted? What practical implications does it have for wilderness preservation, endangered species, ecological restoration, farming, dam building, wildlife management, and hunting and fishing? Are those implications acceptable? Why or why not?
4. What is the eco-fascism objection to the land ethic? Is it a good objection?
5. Do inanimate objects of nature such as rocks, soils, streams, wetlands, deserts, and rainforests have moral standing? Should any be accorded legal rights? Why or why not?
6. Do you agree with the author's suggestion that Leopold's summary moral maxim should be either substantially revised or rejected? Why or why not?

Further Reading

The starting point for any serious study of Leopold's land ethic remains his timeless *A Sand County Almanac with Essays on Conservation from Round River* (New York: Ballantine Books, 1970; originally published in a shorter version by Oxford University Press in 1949). For an extensive collection of Leopold's writings and letters, see Curt Meine, ed., *Aldo Leopold:* A Sand County Almanac *and Other Writings on Ecology and Conservation* (New York: Library of America, 2013). For an excellent biography of Leopold, see Curt Meine, *Aldo Leopold: His Life and Work*, rev. ed. (Madison, WI: University of Wisconsin Press, 2010). Though sometimes unreliable, J. Baird Callicott remains Leopold's foremost interpreter and philosophical disciple. See especially Callicott's edited collection, *A Companion to* A Sand County Almanac*: Interpretive and Critical Essays* (Madison, WI: University of Wisconsin Press, 1987); and his *Thinking Like a Planet: The Land Ethic and the Earth Ethic* (New York: Oxford University Press, 2013). For a short and appreciative overview of Leopold's environmental thought, see Roderick Frazier Nash, *Wilderness and the American Mind*, 5th ed. (New Haven, CT: Yale University Press, 2014), pp. 182–99.

Chapter 6

Deep Ecology and Ecofeminism

Beginning in the 1960s, a number of radical forms of environmental theory emerged and continue to have many followers. The two most influential were **deep ecology** and **ecofeminism**. Let's begin with deep ecology.

6.1 Deep Ecology

The term "deep ecology" was coined by Arne Naess (1912–2009), a distinguished Norwegian philosopher and environmentalist. In a classic 1973 paper,[1] Nacss distinguishcd "dccp" from "shallow" approaches to the environment. Shallow ecology is thoroughly human-centered and is concerned with ecological problems, such as pollution and resource depletion, solely because of their negative impact on human beings. Deep ecology, by contrast, is not anthropocentric but rather life-centered. Deep ecologists claim that all living things have inherent value and that humans are parts of nature, not above or outside it. Because they see environmental problems as having deep causes and pervasive effects, deep ecologists ask radical and searching questions about how we should live, how societies should be structured, and how humans should interact with nature. Many common attitudes and assumptions—that humans are a superior life-form and have a right to dominate the planet, that humans are separate from nature, that governments should promote ever-rising economic growth—are rejected by deep ecologists. In their view, "shallow" approaches to the environment are not sufficient; deep and far-reaching social and economic changes are needed to heal nature, live sustainably, improve life quality, and restore the rightful place of humans in the natural order.

1. Arne Naess, "The Shallow and the Deep, Long-Range Ecology Movement: A Summary," *Inquiry* 16 (1973), pp. 95–100.

Deep ecologists disagree on many issues; they come from all sorts of different backgrounds, faiths, cultural traditions, and ideological commitments. But there is a common credo that most deep ecologists accept.[2] This is known as "The Deep Ecology Platform" or "The Eight Points." Arne Naess and the American philosopher George Sessions formulated the Platform during a camping trip in Death Valley, California in 1984. Slightly different formulations of the Platform exist. Here is the original 1984 version:

1. The well-being and flourishing of human and nonhuman life on Earth have value in themselves (synonyms: intrinsic value, inherent worth). These values are independent of the usefulness of the nonhuman world for human purposes.
2. Richness and diversity of life-forms contribute to the realization of these values and are also values in themselves.
3. Humans have no right to reduce this richness and diversity except to satisfy *vital* needs.
4. The flourishing of human life and cultures is compatible with a substantially smaller human population. The flourishing of nonhuman life requires a smaller human population.
5. Present human interference with the nonhuman world is excessive and the situation is rapidly worsening.
6. Policies must therefore be changed. These policies affect basic economic, technological, and ideological structures. The resulting state of affairs will be deeply different from the present.
7. The ideological change will be mainly that of appreciating *life quality* (dwelling in situations of inherent value) rather than adhering to an increasingly higher standard of living. There will be a profound awareness of the difference between big and great.

---

2. Though by no means all. Some leading theorists believe that the heart of deep ecology is not the Platform, but the brand of metaphysical holism that Naess called "relationism"—the view that reality is constituted by relationships and that nature and self are ultimately one. See, for example, Freya Mathews, "Deep Ecology," in Dale Jamieson, ed., *A Companion to Environmental Philosophy* (Malden, MA: Blackwell, 2001), pp. 224–25; and Warwick Fox, "Deep Ecology: A New Philosophy for Our Time?" *The Ecologist* 14 (1984), pp. 194–200; reprinted in Louis P. Pojman and Paul Pojman, *Environmental Ethics: Readings in Theory and Practice*, 6th ed. (Boston, MA: Wadsworth, 2012), p. 152.

8. Those who subscribe to the foregoing points have an obligation directly or indirectly to try to implement the necessary changes.[3]

To gain a better understanding of deep ecology, let's look briefly at each of these eight points in turn.

1. The well-being and flourishing of human and nonhuman life on Earth have value in themselves (synonyms: intrinsic value, inherent worth). These values are independent of the usefulness of the nonhuman world for human purposes.

This principle expresses a clear rejection of anthropocentrism—or more precisely of extreme forms of anthropocentrism that claim that only human beings have intrinsic value or moral standing. All forms of life, both plants and animals, are claimed to have value or worth that is inherent in the sense of being both for its own sake and not being dependent on "any awareness, interest, or appreciation of it by any conscious being."[4] In commenting on Point 1, Naess notes that "life" is used broadly to include collective entities, such as species and ecosystems, as well as nonliving things such as rivers, mountains, and landscapes. Deep ecology is thus a form of ecocentrism, since it recognizes that both living and nonliving natural objects and ecological wholes have moral standing. This moral standing, moreover, is considerable, calling for a significant degree of moral respect. As we saw earlier, many anthropocentrists would agree that all living things have *some* intrinsic value, while claiming that humans have vastly (perhaps incomparably) more. Deep ecologists, by contrast, claim that all living things have significant moral status, and thus deserve "fundamental concern and respect."[5] Naess himself offers no argument for his claim that all life-forms have substantial intrinsic value, claiming that it rests on "intuition" and is incapable of proof.[6]

3. Bill Devall and George Sessions, *Deep Ecology: Living as if Nature Mattered* (Salt Lake City, UT: Gibbs Smith, 1985), p. 70.

4. Arne Naess, "The Deep Ecological Movement: Some Philosophical Aspects," *Philosophical Inquiry* 8 (1986), pp. 10–31; reprinted in George Sessions, ed., *Deep Ecology for the 21st Century: Readings on the Philosophy and Practice of the New Environmentalism* (Boston, MA: Shambala Publications, 1995), p. 69. The phrase in quotation marks is a quote from Tom Regan.

5. Naess, "The Deep Ecological Movement," p. 68. As we shall see, Naess was a biocentric egalitarian, who believed that all living things have equal moral standing. This was part of his personal environmental philosophy, but he did not include this egalitarian claim in the Platform because he believed it was too controversial.

6. Arne Naess, *Life's Philosophy: Reason and Feeling in a Deeper World*, translated by Roland Huntford (Athens, GA: University of Georgia Press, 2002), p. 65. See also Stephen Bodian, "Simple in Means, Rich in Ends: An Interview with Arne Naess," in Sessions, *Deep Ecology for the 21st Century*, p. 28.

2. Richness and diversity of life-forms contribute to the realization of these values and are also values in themselves.

Naess, like all deep ecologists, believed strongly in the importance of preserving biodiversity and the wild habitats necessary to support it. According to deep ecologists, even "lower" or "simple" organisms have value, both in themselves and by contributing to the flourishing and well-being of other living things. They should not be viewed as "merely steps toward the so-called higher or rational life forms."[7]

3. Humans have no right to reduce this richness and diversity except to satisfy vital *needs.*

Here we see clearly the deep and far-reaching changes that deep ecologists call for. In opposition to all shallow, reformist, or mainstream environmentalism, deep ecologists deny that humans have a right to dominate the planet and heavily prioritize their own interests over those of other living creatures. They claim that if nonhuman creatures are to be treated with the "fundamental concern and respect" they deserve, humans must dramatically scale back their domineering and destructive footprint on the planet. Humans may not reduce the abundance or diversity of earthly life-forms except to satisfy vital human needs. What counts as "vital"? Naess notes that the Platform leaves the term deliberately vague to allow for a wide range of views.[8] He explains, however, that what counts as "vital" will vary from culture to culture,[9] and that they include not only basic survival needs but also necessities that give "life its deepest meaning."[10] Naess claims that whenever the non-vital needs of humans "come into conflict with the vital needs of nonhumans, then humans should defer to the latter."[11]

4. The flourishing of human life and cultures is compatible with a substantially smaller human population. The flourishing of nonhuman life requires a smaller human population.

Deep ecologists believe that many of today's most urgent environmental problems—climate change, species extinction, resource depletion, air and water

7. Naess, "The Deep Ecological Movement," p. 69.
8. Naess, "The Deep Ecological Movement," p. 69.
9. Naess, "The Deep Ecological Movement," p. 69.
10. Naess, *Life's Philosophy*, p. 107.
11. Naess, "The Deep Ecological Movement," p. 74.

pollution, deforestation, overconsumption, overfishing, to name a few—are directly related to human overpopulation. In 1800, there were approximately a billion human beings on the planet. Today there are 7.8 billion. With a rising global standard of living, such high numbers inevitably take a heavy toll on the planet and its nonhuman inhabitants. If all living things have "an equal right to live and blossom,"[12] as deep ecologists believe, we must recognize that the human population must shrink. What level of human population would be ideal? Naess once floated the figure of one hundred million.[13] Deep ecologists stress that such a drastic reduction in human population would likely take centuries, and must be achieved through humane, non-coercive methods.[14] At the same time, the pace of climate change, species extinction, and environmental destruction demands that we address overpopulation as an immediate and urgent concern.

5. Present human interference with the nonhuman world is excessive, and the situation is rapidly worsening.

Naess characterized this statement as "mild" in the mid-1980s,[15] and most environmentalists would use stronger language today. (Terms such as "ecocide" and "ecocatastrophe" are now common in environmental discourse.) By most measures of environmental health, including climate change, biodiversity loss, deforestation, and resource depletion, human "interference" with nature is unquestionably a good deal worse than it was when the Deep Ecology Platform was written in 1984.

---

12. Arne Naess, "The Shallow and the Deep, Long-Range Ecology Movement," pp. 95–100; reprinted in Alan Drengson and Yuichi Inoue, eds., *The Deep Ecology Movement: An Introductory Anthology* (Berkeley, CA: North Atlantic Books, 1995), p. 4. Naess claims that this norm should be taken as a "guideline" to behavior, rather than as a moral absolute (ibid., p. 167). He recognizes, for example, that in conflict situations, humans are often justified in giving priority based on "nearness" of social ties and relationships. See Arne Naess, "Equality, Sameness, and Rights," in Sessions, *Deep Ecology for the 21st Century*, p. 222.

13. Arne Naess, *Ecology, Community, and Lifestyle: Outline of an Ecosophy*, translated and edited by David Rothenberg (Cambridge: Cambridge University Press, 1989), pp. 140–41.

14. Naess, "The Deep Ecology Movement," p. 69.

15. Naess, "The Deep Ecology Movement," p. 69.

6. Policies must therefore be changed. These policies affect basic economic, technological, and ideological structures. The resulting state of affairs will be deeply different from the present.

In contrast to shallow or reform environmentalism, deep ecology calls for fundamental and wide-ranging changes in how humans live and relate to nature. How radical? On this point the Platform is unhelpfully vague and tends to mask the truly revolutionary (and often utopian) character of much deep ecological thinking. When one delves into the writings of Naess and other leading deep ecologists, it becomes clear how strongly they were affected by the leftist counterculture of the 1960s and 1970s. Naess, for example, envisions a future of small, decentralized, mostly self-sufficient, nature-friendly "Green societies" that reject economic growth, free-market capitalism, consumerism, class hierarchies, large-scale industry, and sophisticated technology, while embracing cultural diversity, participatory democracy, vegetarianism, nonviolence, and a standard of living that is "not much different from and higher than the needy."[16] The scale and dreamy idealistic flavor of such proposed changes have led some to characterize deep ecology as the "radical fringe"[17] of the environmental movement.

7. The ideological change will be mainly that of appreciating life quality *(dwelling in situations of inherent value) rather than adhering to an increasingly higher standard of living. There will be a profound awareness of the difference between big and great.*

This point reflects deep ecology's commitment to sustainable green lifestyles that are, in Naess's oft-quoted phrase, "simple in means, rich in ends." The focus of such lifestyles will be on "activities most directly serving values in themselves and having intrinsic value,"[18] rather than on pursuing material affluence or other instrumental ends. No definition of "life quality" is offered in the Platform, but Naess elsewhere suggests that a high-quality life is one in which a person is able to live the kind of life that, after careful reflection, they would really wish to live.[19] Following Mahatma Gandhi (1869–1948), whose thought he studied closely, Naess personally believed that the goal of human life is to achieve "self-realization," which he held involved both fulfillment of one's

16. Naess, "Deep Ecology and Lifestyle," in Sessions, *Deep Ecology for the 21st Century*, p. 260. See also Naess, *Ecology, Community, and Lifestyle*, pp. 144–55.

17. Charles T. Rubin, *The Green Crusade: Rethinking the Roots of Environmentalism* (New York: Free Press, 1994), p. 27.

18. Naess, "Deep Ecology and Lifestyle," p. 259.

19. Naess, *Ecology, Community, and Lifestyle*, p. 123.

inner potential and recognition of one's ultimate oneness with Nature or Ultimate Reality.[20] These notions of self-realization and pantheistic non-duality are facets of Naess's personal philosophy, which he called Ecosophy T, and not defining features of deep ecology.

8. Those who subscribe to the foregoing points have an obligation directly or indirectly to try to implement the necessary changes.

Among the hallmarks of deep ecology is a strong commitment to ecological activism. Naess himself participated in environmental protests and was once arrested for doing so.[21] Although a number of militant environmental groups, including Earth First! and the Earth Liberation Front, have drawn inspiration from deep ecology, Naess and most other leading figures in the movement were firmly committed to legal and nonviolent forms of environmental activism and protest.[22]

6.2 The Future of Deep Ecology

The palmy days of deep ecology were in the period from the mid-1970s to the mid-1990s. Is it a spent force today? Though deep ecology has clearly lost some of its vitality as a movement, I do not believe that it is dead. Some of the core themes of deep ecology—most notably, the intrinsic value of nature, the vital importance of preserving wilderness and biodiversity, and the need for ecological activism—have been absorbed into mainstream environmental thought.[23] Moreover, many environmentalists today would continue to agree with much of the spirit of the Deep Ecology Platform. The only major points of disagreement would be over Point 3 (the claim that only vital human needs justify interference with nature) and Point 4 (the call for a substantially smaller human population). To some extent, therefore, deep ecology can be seen as a

20. Naess, "Self-realization: An Ecological Approach to Being in the World," in Sessions, *Deep Ecology for the 21st Century*, pp. 225–39; and Naess, *Ecology, Community, and Lifestyle*, pp. 196–210.

21. Naess's arrest is recounted in the 1997 documentary *The Call of the Mountain: Arne Naess and the Deep Ecology Movement*, YouTube. Web. 24 February 2020.

22. Naess, "Deep Ecology and Lifestyle," p. 261; and Devall and Sessions, *Deep Ecology*, p. 198.

23. See, for example, Pope Francis's 2015 encyclical on the environment, *Laudato Si'* (widely available online), where each of these themes is clearly sounded.

victim of its own success. Its message was so widely embraced that it may now appear passé.

In reflecting on the continued vibrancy of deep ecology as a movement, it is important not to confuse the personal environmental philosophies of Naess and other leading deep ecologists with deep ecology itself. On a personal level, as we have seen, Naess believed in the ultimate oneness of all reality and embraced biocentric egalitarianism, the belief that all living things have equal inherent value. Yet neither of these views is expressed in the Platform. This is also the case with Naess's doctrine of self-realization and his hankering for small, classless, low-tech, anti-capitalist, deeply democratic Green communities. Many of those views also may not be widely popular today, but they are aspects of Naess's personal environmental philosophy (Ecosophy T), not of deep ecology itself.[24]

It is certainly true that deep ecology drew much of its energy from the 1960s and 1970s counterculture; a period when Eastern mysticism, hippie communes, civil disobedience, attacks on "squares" and "the Man," and back-to-nature pastoralism were all the rage. It was also a time when people generally were more hopeful about the possibility of achieving fundamental and transformative social change. The early deep ecologists were optimists in that sense. They believed they could convince large numbers of people to radically alter their lifestyles for the sake of ecological health and a Green future. That may seem rather utopian today, but it remains to be seen how climate change will affect future lifestyles, work, and social and political institutions. It is possible that "basic economic, technological, and ideological structures" will indeed change, just as the deep ecologists urged. Should such fundamental changes involve *gender relations* as well as other basic social relations and institutions? Another prominent group of radical environmentalists—ecofeminists—argue powerfully that they must and should. To ecofeminism we now turn.

24. Naess produced a famous illustration, the Apron Diagram, to represent the various levels of the deep ecology movement. The diagram includes the Platform (Level 2); the personal philosophical, religious, or ideological reasons why people support the Platform (Level 1); the general consequences and policies that flow from the Platform (Level 3); and the level of specific implications and concrete decisions (Level 4). According to Naess, supporters of deep ecology may hold differing views at all levels except Level 2, the Platform. Arne Naess, "The Apron Diagram," in Drengson and Inoue, *The Deep Ecology Movement*, pp. 11–12.

6.3 Ecofeminism

Ecofeminism is a family of radical environmental theories that emerged in the 1970s and 1980s as an outgrowth of the environmental and women's liberation movements. Like feminism itself, ecofeminism comes in many different forms. Some ecofeminists, for example, embrace earth-centered forms of spirituality, while others may be deep ecologists, Marxists, liberals, or critical theorists. What all ecofeminists have in common is the threefold claim that (1) there are important connections between the domination of women and the domination of nature, (2) understanding those connections is crucial to both feminism and an adequate environmental ethic, and (3) both forms of domination must be opposed and ended.

What sorts of important connections might there be between sexism and naturism? One leading contemporary ecofeminist, Karen J. Warren, helpfully explores ten asserted linkages.[25] Here let's focus on the two most significant alleged connections: historical and conceptual.

According to ecofeminists, sexism and naturism are linked historically because both have similar causes. Under patriarchy, for example, nature has been feminized (as in "Mother Earth" or "Mother Nature") and women have been naturized (as in such derogatory epithets as "foxes," "chicks," "vixens," "sex kittens," "cougars," and "bitches"). According to many ecofeminists, such equations, when combined with the sexist assumption that men are superior to women, have contributed greatly to the domination of both women and nature.

Ecofeminists also claim that sexism and naturism are conceptually linked because attempts to justify sexism and naturism typically depend on a similar kind of faulty logic. According to Karen Warren, two conceptual linkages are especially important. One is what she calls hierarchically organized value dualisms—pairs of contrasted concepts in which one disjunct is valued more highly than the other. Common value dualisms in Western civilization have included reason/emotion, mind/body, human/nature, and man/woman. Because women historically have been associated with the supposedly inferior qualities of emotion, body, and nature, such dualisms often figure, explicitly or implicitly, in rationalizations for sexism. And because humans historically have been viewed as above and separate from nature, they also have featured in arguments for naturism.

According to Warren, there is another major conceptual connection between sexism and naturism. Both depend on what she terms a **logic of domination**—the false and oppressive assumption that superiors have a right to

---

25. Karen J. Warren, *Ecofeminist Philosophy: A Western Perspective on What it Is and Why it Matters* (Lanham, MD: Rowman & Littlefield, 2000), pp. 21–38.

dominate, exploit, or subordinate inferiors. Warren argues that because feminists oppose sexism, and sexism and naturism both rest on a similar logic of domination, feminists should also oppose naturism.

To flesh out these claims, Warren formulates an argument for critical analysis that makes explicit how value dualisms and a logic of domination have played a crucial role in arguments for patriarchy. Her argument runs as follows:

> (B1) Women are identified with nature and the realm of the physical; men are identified with the "human" and the realm of the mental.
>
> (B2) Whatever is identified with nature and the realm of the physical is inferior to ("below") whatever is identified with the "human" and the realm of the mental: or, conversely, the latter is superior to ("above") the former.
>
> (B3) Thus, women are inferior to ("below") men; or, conversely, men are superior to ("above") women.
>
> (B4) For any X and Y, if X is superior to Y, then X is justified in subordinating Y.
>
> (B5) Thus, men are justified in subordinating women.[26]

In this argument, both value dualisms (in premises B1–B3) and the logic of domination (in premise B4) are used to rationalize the subordination of women by men.

Curiously, Warren claims that all feminists must oppose the logic of domination because all feminists oppose patriarchy (B5), and patriarchy "rests" on the logic of domination.[27] This is a slip. Some feminists, in principle, might accept the logic of domination but avoid patriarchy simply by denying that women are inferior to men. To show that all feminists must oppose the logic of domination Warren must show that *any* plausible argument for patriarchy must depend on the logic of domination, and this she fails to do. Historically, of course, some arguments for patriarchy have depended not on assumptions of male superiority but on beliefs that men and women are just *different* in ways that justify differential gender roles.[28]

---

26. Karen J. Warren, "The Power and Promise of Ecological Feminism," *Environmental Ethics* 12:3 (1990), pp. 125–46; reprinted in Karen J. Warren, ed., *Ecological Feminist Philosophies* (Bloomington, IN: Indiana University Press, 1996), p. 22.

27. Warren, "The Power and Promise of Ecological Feminism," p. 23.

28. Margarita Garcia Levin and Michael Levin, "A Critique of Ecofeminism," in Louis P. Pojman, ed., *Environmental Ethics: Readings in Theory and Application*, 3rd ed. (Belmont, CA: Wadsworth, 2001), pp. 199–204.

MAKING A DIFFERENCE

Vandana Shiva: A Voice for the Victims

Globalization has raised standards of living around the world and enabled over a billion people to lift themselves out of extreme poverty over the past quarter-century. But globalization produces losers as well as winners. Today, one of the most prominent voices for the left-out is Indian author and activist Vandana Shiva (1952–).

Shiva was born in Dehradun, India, in the foothills of the Himalaya Mountains, and educated in both India and Canada, receiving a PhD in the philosophy of physics from the University of Western Ontario in 1978. She first became involved with ecological activism when she volunteered for the Chipko Movement in the mid-1970s. In that much-publicized protest action, rural villagers—mostly women—succeeded in stopping rampant deforestation by hugging trees in acts of nonviolent civil disobedience.

Author of more than twenty books, Shiva has become one of the world's best-known critics of genetically modified foods, unchecked global capitalism, economic colonialism, gender inequality, corporate control of the food supply, inequitable distributions of water resources, and replacement of traditional organic agriculture by fertilizer-intensive cash-crop monocultures.

In 1993, Shiva coauthored a pioneering book on ecofeminism with the German feminist and activist Maria Mies.

For her work on behalf of women, nature, and the poor, Shiva has won dozens of awards, including the Right Livelihood Award (the so-called Alternative Nobel Prize) and the Sidney Peace Prize.

Much of Shiva's work has been directed at exposing the dark side of globalized capitalism, noting, for example, how industrial development funded by multinational corporations can drive poor people from their lands, increase food insecurity, erase cultural diversity, widen the gap between rich and poor, corrupt democratic politics, and fuel environmental degradation. As Shiva pointedly remarks, "nature shrinks as capital grows."[29]

29. Quoted in Howard Chua-Eoan, "Vandana Shiva: Prophet of Boom and Doom," *Time* magazine, November 18, 2013. Web. 14 July 2020.

Further, it's not clear that anyone would accept the logic of domination in the bald, categorical way Warren formulates it ("For any X and Y, if X is superior to Y, then X is justified in subordinating Y"). No one would claim, for example, that because Billy is better at fourth-grade math than Johnny, Billy is entitled to "subordinate" Johnny. It seems obvious that some kinds of superiority justify hierarchical structures and some don't. Would Warren deny that parents have a right (and duty) to direct the upbringing of their children? Or that teachers, in virtue of their superior knowledge and educational attainments, have a right to manage their classrooms and direct the education of their pupils? It simply invites confusion to discuss "a logic of domination" in the categorical, all-or-nothing way Warren does.

Such confusion is evident in Warren's attempt to show that feminists must oppose *both* naturism and sexism. She argues:

> (C1) Feminism is a movement to end sexism.
>
> (C2) But sexism is conceptually linked with naturism (through an oppressive conceptual framework characterized by a logic of domination).
>
> (C3) Thus, feminism is (also) a movement to end naturism.[30]

This argument assumes that superiority never justifies hierarchy or subordination, which we've seen is false. We must thus ask: Are there morally relevant differences between the subordination of women by men and the subordination of nature by humans? If so, then naturism might be justifiable while sexism is not.

And that, of course, is what a lot of people believe. To belabor the obvious: Sexism is a form of invidious discrimination that causes pain, resentment, disempowerment, humiliation, and a host of other injuries and ills. Naturism sometimes produces these or other negative consequences but often it does not. As Simon and Garfunkel sing, "a rock feels no pain, and an island never cries."[31] Indeed, as environmental philosopher Frederik Kaufman points out, it's not clear that a rock formation, say, can be literally "dominated" or "oppressed" in any meaningful way, since this seems to presuppose some sort of inner drive or propensity in the rocks that is being frustrated.[32] Thus, even if it is true that

---

30. Warren, "The Power and Promise of Ecological Feminism," p. 25.

31. Paul Simon and Art Garfunkel, "I Am a Rock," LyricFind. Web. 7 March 2020.

32. Frederik A. Kaufman, *Foundations of Environmental Philosophy: A Text with Readings* (New York: McGraw-Hill, 2003), p. 389.

sexism and naturism are linked in some important ways, it doesn't follow that they are somehow on a moral par or that they should both be rejected.

Warren, in fact, like many ecofeminists, seems to exaggerate the connections between sexism and naturism. As Huey-li Li notes, many cultures (such as ancient China) have combined respect for nature with markedly sexist attitudes toward women.[33] Conversely, some cultures (such as Soviet-dominated Eastern Europe) have had poor environmental records while (nominally at least) being committed to gender equality. To say that patriarchy is *the* cause of the environmental crisis, or even a major cause, is highly doubtful.

One final problem with ecofeminism should be noted. Ecofeminists rarely spell out in detail what sorts of environmental practices and policies they support. Warren, for example, never explains what she means by "nature" or what counts as unjustifiable "oppression" or "subordination" of nature. Is it permissible, for example, for humans to build dams, engage in commercial fishing, weed lawns, cull invasive species, cut forests for lumber, keep pets, experiment on animals, or drill for oil in the Arctic? As Margarita and Michael Levin note, it is difficult to evaluate ecofeminism without knowing its "cash value"—that is, what it does and does not concretely permit in our treatment of the environment.[34]

Despite such problems, the contributions of leading ecofeminist thinkers cannot be denied. Ecofeminists have made permanent contributions to environmental thought by pointing out connections, both historical and conceptual, between sexism and naturism, and by calling into question oppressive frameworks that often undergird indefensible and discriminatory hierarchies and value dualisms.

Chapter Summary

1. This chapter explored two radical environmental philosophies: deep ecology and ecofeminism.
2. Deep ecologists are opposed to mainstream ("shallow") ecology. They call for much deeper and more fundamental changes in our environmental practices and attitudes toward nature than do mainstream ecologists.

33. Huey-li Li, "A Cross-Cultural Critique of Ecofeminism," in Greta Gaard, ed., *Ecofeminism: Women, Animals, Nature* (Philadelphia, PA: Temple University Press, 1993), p. 276.

34. Levin and Levin, "A Critique of Ecofeminism," p. 202.

3. Though deep ecologists disagree about many issues, most embrace the eight principles expressed in the Deep Ecology Platform (co-written by Arne Naess and George Sessions). These principles affirm the inherent value of all living things and the importance of preserving biodiversity, call for a substantial reduction in human population, assert the need for fundamental economic and technological changes to protect nature, emphasize life-quality over economic growth and material affluence, and strongly endorse environmental activism and engagement.
4. Ecofeminists believe that there are significant connections between sexism (the systematic subordination of women) and naturism (the systematic subordination of nature). According to ecofeminist scholar Karen J. Warren, both arise from a faulty and oppressive logic of domination that assumes that superiors have a right to dominate inferiors.

Discussion Questions

1. What is deep ecology? How does deep ecology differ from shallow ecology? Why is deep ecology often seen as a radical environmental ethic?
2. Which points, if any, in the Deep Ecology Platform do you disagree with? Why?
3. Some deep ecologists, like Arne Naess, are mystics or pantheists in the sense that they believe that there is one ultimate reality, God, and that humans are ultimately one with God and Nature. Do you agree with this mystical view of reality? What reasons or grounds might there be for believing it? Generally speaking, does such mysticism support environmentally friendly actions?
4. Do you agree with Naess's oft-quoted claim that people should pursue lifestyles that are "simple in means but rich in ends?" Why or why not?
5. Do you believe that climate change might lead to a revival of deep ecology? Why or why not?
6. What is ecofeminism? Do you agree with its central claims?
7. What historical connections are there between sexism and naturism? Are such connections significant or relatively superficial?

8. How, according to Warren, are sexism and naturism conceptually linked? Does Warren succeed in showing that feminists must consistently oppose naturism as well as sexism?

Further Reading

For Naess's writings on deep ecology, see his *The Ecology of Wisdom: Writings by Arne Naess*, edited by Alan Drengson and Bill Devall (Berkeley, CA: Counterpoint, 2008); *Ecology, Community, and Lifestyle: Outline of an Ecosophy*, translated and edited by David Rothenberg (Cambridge: Cambridge University Press, 1989); and *Deep Ecology of Wisdom*, edited by Harold Glasser and Alan Drengson (Dordrecht, the Netherlands: Springer, 2005). Some of Naess's writings can be difficult for readers new to philosophy. For more accessible introductions to deep ecology, see Bill Devall and George Sessions, *Deep Ecology: Living as if Nature Mattered* (Salt Lake City, UT: Gibbs Smith, 1985); and Bill Devall, *Simple in Means, Rich in Ends: Practicing Deep Ecology* (Salt Lake City, UT: Gibbs Smith, 1988). For a valuable and extensive collection of readings on deep ecology, see George Sessions, ed., *Deep Ecology for the 21st Century: Readings on the Philosophy and Practice of the New Environmentalism* (Boston, MA: Shambala Publications, 1995). For critical reactions to deep ecology, see Eric Katz, Andrew Light, and David Rothenberg, eds., *Beneath the Surface: Critical Essays in the Philosophy of Deep Ecology* (Boston, MA: MIT Press, 2000). For a lucid, sympathetic, yet critical overview of deep ecology, see Freya Mathews, "Deep Ecology," in Dale Jamieson, ed., *A Companion to Environmental Philosophy* (Malden, MA: Blackwell, 2001), pp. 218–32. For a classic analysis of the connections between women and nature in Western thought, see Carolyn Merchant, *The Death of Nature: Women, Ecology, and the Scientific Revolution* (New York: Harper and Row, 1983). For a useful collection of essays on ecofeminism, see Karen J. Warren, ed., *Ecological Feminist Philosophies* (Bloomington, IN: Indiana University Press, 1996). Also worth consulting is Warren's monograph, *Ecofeminist Philosophy: A Western Perspective on What it Is and Why it Matters* (Lanham, MD: Rowman & Littlefield, 2000). For a defense of Warren's ecofeminist view and a response to her critics, see Amy L. Goff-Yates, "Karen Warren and the Logic of Domination: A Defense," *Environmental Ethics* 22:2 (2000), pp. 169–81.

Chapter 7

Moderate Ecocentrism

In previous chapters, we looked at a number of leading theories of environmental ethics. We noted that each confronts substantial objections. Now we must ask: What would a satisfactory environmental ethic be like?

Two lessons have emerged clearly from previous chapters. First, a sound environmental ethic should be balanced and grounded in a firm sense of reality. For decades, environmental ethics has been dominated by views rooted in 1960s radicalism. To claim that all living things have equal moral status (Paul Taylor), or that all animal testing should be abolished (Tom Regan), or that humans may interfere with nature only in order to satisfy vital needs (Arne Naess and George Sessions), or that we should reduce human population by over 95 percent (Naess) is fantasy, not serious philosophy. We face a planetary emergency, a critical moment in Earth's history, and what is urgently needed now is an environmental ethic that can serve as a big tent within which people of many different faiths and backgrounds can unite and work together to heal the earth and secure a Green tomorrow.

Second, a satisfactory environmental ethic must reject any form of extreme or tyrannical anthropocentrism. Any view that claims that only human interests matter, or that we have a right to overrun and trash the planet, is both false and destructive. We must recognize that humans are latecomers on this ancient earth who should respect nature, learn to live in harmony with it, and work to preserve its beauties and resources for generations to come.

With these two lessons in mind, I wish to sketch a form of **moderate ecocentrism** that I believe fits the bill. As befits an introductory text like this, my discussion will be brief and programmatic. A much more in-depth discussion would be needed to fully develop the theory and defend its central claims.

Like Naess and Sessions in their presentation of the Deep Ecology Platform (see Chapter 6), I will state my theory as a series of general principles, followed by some explanatory comments.

Moderate Ecocentrism: Basic Principles

1. All living things have inherent worth and deserve moral respect and consideration.
2. Some living things have greater inherent worth than others. Human beings, in particular, have a special dignity and inherent worth. As a result, human interests should generally take precedence over those of other life-forms.
3. Some ecological wholes, such as ecosystems and species, have inherent worth and deserve moral respect and consideration.
4. In environmental matters, our primary concern should generally be for the health and well-being of ecological wholes, such as species and ecosystems, rather than for the good of individual organisms.

Comments

Comments on Principle 1: By "inherent worth" I mean intrinsic value—that is, the value a thing has in itself or for its own sake, apart from any use or instrumental value it may have to humans or other organisms.[1] Many plants and animals, of course, have use-value to humans. Honeybees, for example, play a vital role in pollinating crops, and microbes in our intestines help us digest our food. But aside from their economic or other instrumental value, living things also have value in themselves. This is the case, in part, because as Paul Taylor and Lawrence E. Johnson argue, all living things are goal-oriented centers of life that strive to preserve themselves and realize their own good in their own unique way.[2] Each living thing has a good of its own, a distinctive mode of flourishing, a way of being-well and faring-well in its biological niche. Put otherwise, all organisms have well-being interests—stakes in the way their lives play out. A dog, for example, has a well-being interest in obtaining food, and a potted plant has a well-being interest in being watered. Contrary to David Schmidtz,[3] even noxious and annoying organisms like mosquitoes have

---

1. Michael J. Zimmerman and Ben Bradley, "Intrinsic vs. Extrinsic Value," *The Stanford Encyclopedia of Philosophy*. Web. 13 May 2020.

2. Paul W. Taylor, *Respect for Nature: A Theory of Environmental Ethics* (Princeton, NJ: Princeton University Press, 1986), p. 121; and Lawrence E. Johnson, *A Morally Deep World: An Essay on Moral Significance and Environmental Ethics* (New York: Cambridge University Press, 1991), p. 134.

3. David Schmidtz, "Are All Species Equal?" *Journal of Applied Philosophy* 15 (1998), pp. 57–67; reprinted in David Schmidtz and Elizabeth Willott, eds., *Environmental*

intrinsic value and deserve some degree of respect. Though a world without mosquitoes would be a much better world for humans and numerous other animals, mosquitoes are nonetheless marvelously complex miracles of evolution and have welfare-interests and a good of their own. Thus, they have inherent worth.

This inherent worth, it should be noted, is objective and does not depend, as some theorists claim,[4] on human feelings, beliefs, or valuations. Values do not necessarily require valuers. A ruby-throated hummingbird does not suddenly acquire intrinsic value the moment it first flashes before an appreciative human eye; such a claim is baldly anthropocentric. If the human race were on the brink of extinction, and the last surviving human could with his last breath painlessly exterminate all life on earth, would this be morally wrong?[5] If values cannot exist without valuers, it seems that nothing of normative consequence would be lost. Yet intuitively we feel that such a deed would be an act of wanton and monstrous evil. If so, this suggests—though admittedly it does not prove—that there are real, objective values in nature that are not dependent on human or other sentient valuers or perceivers.[6]

Comments on Principle 2: As we saw in previous chapters, some environmental ethicists claim that all living things have equal inherent worth (Taylor) or at least that all sentient or higher animals do (Singer and Regan). Such egalitarian claims are implausible. Values are not free-floating occult entities; they emerge from, and depend upon, natural or physical features of the world. If a novel, for example, possesses the normative quality of literary greatness, that quality must depend on certain value-conferring qualities of the story (emotional power, psychological depth, trueness-to-life, etc.). Thus, if something in nature has

---

Ethics: What Really Matters, What Really Works, 2nd ed. (New York: Oxford University Press, 2012), p. 120.

4. See Dale Jamieson, *Ethics and the Environment: An Introduction* (New York: Cambridge University Press, 2008), pp. 74–75; and J. Baird Callicott, *In Defense of the Land Ethic: Essays in Environmental Philosophy* (Albany, NY: State University of New York Press, 1989), pp. 133–34.

5. This is a version of the so-called last man argument, first proposed by Richard Routley in "Is There a Need for a New, an Environmental Ethic?" *Proceedings of the XVth World Conference of Philosophy* (Varna, Bulgaria, 1973), 1/6, pp. 205–10; reprinted in J. Baird Callicott and Robert Frodeman, eds., *Encyclopedia of Environmental Ethics and Philosophy*, vol. 2 (Farmington Hills, MI: Macmillan Reference, 2009), p. 487. For a critical response, see Jamieson, *Ethics and the Environment*, pp. 74–75.

6. For more on why values in nature are not entirely subjective or mind-dependent, see Holmes Rolston III, *Philosophy Gone Wild* (Buffalo, NY: Prometheus Books, 1989), pp. 91–117.

inherent value, it must be in virtue of certain physical features or other attributes that ground such value. So what sorts of properties do confer intrinsic value on organisms or ecological wholes?

We have seen that *having-a-good-of-its-own* is one property that grounds intrinsic value, and that it is this attribute that gives all living things at least some degree of inherent worth. But having-a-good-of-its-own is by no means the only attribute that confers intrinsic value on a being or an ecological collective; qualities like beauty, biodiversity, generativity, sentience, rationality, knowledge, autonomy, moral responsiveness, emotional capacity, aesthetic sensibility, and a capacity for love and happiness are, and ought to be, valued for their own sakes. And since some things in nature possess those qualities (in varying degrees) and others do not, it follows that some organisms or other natural objects have greater intrinsic value or inherent worth than others.

As environmental philosopher David Schmidtz points out, intrinsic value comes in degrees and can be additive.[7] For example, all other things being equal, a short story that possesses positive literary quality X (e.g., trueness-to-life) is not as good as one that possesses the same degree of X plus positive literary quality Y (e.g., psychological depth). In the same way, an organism that possesses rationality in addition to, say, a capacity to grow and reproduce, has more intrinsic value than an organism that possesses only the ability to grow and reproduce.[8]

By rejecting the biocentric egalitarian's claim that all living things have equal intrinsic value we are able to make commonsense distinctions that fit with widely shared environmental values. For example, we can favor chimpanzees over carrots, and redwoods over the pathogens that threaten them.

7. Schmidtz, "Are All Species Equal?" p. 116. Cf. Aristotle, *On the Generation of Animals*, 731b.

8. Such a view naturally implies that some humans possess greater inherent worth than others. This implication troubles ethicists such as Tom Regan, who believe that it opens the door to elitism, discrimination, and the denial of equal human rights. See Tom Regan, *The Case for Animal Rights* (Berkeley, CA: University of California Press, 1983), pp. 236–37 (arguing that justice requires that we attribute equal inherent worth to all moral agents). But recognizing that people differ in terms of knowledge, intelligence, moral sensitivity, and other factors that contribute to inherent worth does not imply that there are no shared fundamental moral rights. For perhaps all humans (and no nonhumans) equally possess *a certain kind of inherent worth*—sometimes called "dignity"—that entitles them to equal fundamental rights and equal moral respect and concern. See, for example, Pope Francis, *Laudato Si'* (2015), §§33, 42, 119 (available online) arguing that while human beings have "a particular dignity above other creatures," all creatures have "value in themselves" and should be "loved and cherished." Which properties ground such equal inherent dignity is one of the deepest problems of secular and theological ethics.

We can also—in environmentally responsible ways—clear forests, dam rivers, build roads, plant crops, and do other things necessary to build and sustain an advanced civilization and a high quality of human life. For contrary to Aldo Leopold's much-quoted dictum that humans are just "plain citizens" of earth's biotic community, we are in fact much more. Humans, uniquely so far as we know, are creators of many high forms of value. Without us, there would be no scientific understanding, no philosophy, no moral struggle, no artistic triumph, no poetic grandeur, and no spiritual longing. And perhaps one day it will be humans—hitherto the terror and scourge of the natural world—who, through our scientific know-how, will save all life on earth from final extinction.

Comments on Principles 3 and 4: Like all forms of ecocentrism, moderate ecocentrism claims that our primary focus of environmental concern should be with the health and well-being of ecological wholes, rather than with the good of individual plants and animals. This does not mean that we should *never* favor individuals over wholes. It might make sense, for example, to rescue a drowning tiger, even if the tiger (let's assume) has a slightly negative effect on the health of its forest environment. But as a rule, ecocentrists maintain, we should embrace *ecological holism*—the view that, in environmental decision-making, our focal concern should be with the good of ecological wholes.[9] This makes sense for two reasons.

First, ecological wholes often have greater *value* than do individual organisms. In part, this is a matter of sheer numbers. A mountain meadow that contains an abundance of colorful wildflowers has higher intrinsic value than an otherwise identical meadow that has only a few such flowers. The loss of an entire forest, which is home to many species of plants and animals, is a greater loss than that of an individual tree in the forest. Besides numbers, time can also be a factor in determinations of intrinsic natural value. For example, we rightly value an old-growth forest more than we value an individual owl that dwells there, in part because of the much greater antiquity of the forest. But there are also deeper, less obvious reasons why environmental wholes often have more value than individual organisms do. The death of a single blue whale would be sad. The loss of the *species* of blue whales would be an unspeakable tragedy. When we lose a species, we lose forever a pattern, an ontological template. This is why, as Holmes Rolston notes, human-caused species extinction amounts to

9. The terms "ecological holism" or "environmental holism" are sometimes used in alternative senses. For a useful taxonomy, see Don E. Marietta Jr., "Environmental Holism and the Individual," *Environmental Ethics* 10:3 (Fall 1988), pp. 251–58.

a kind of "superkilling."[10] Just as it would be a much greater crime to destroy all remaining copies of Shakespeare's *Macbeth* than it would be to destroy an individual copy, so it would be far worse to cause the extinction of all Bengal tigers than to kill a single member of that species. For similar reasons, destroying an entire ecosystem is also a form of superkilling. For as Rolston reminds us, ecosystems are the engines of evolution, the wombs of life.[11] To destroy an ecosystem is to kill the goose that lays the golden egg.[12] In terms of biotic generativity, a parking lot is a poor substitute for a tropical rainforest.

A second reason why environmentalists should focus mainly on the good of ecological wholes is that it usually simplifies environmental policymaking and produces better outcomes. Numerous problems can arise when we focus on the welfare of individual plants or animals and lose sight of the good of whole ecosystems or species. Which organisms should we favor—lions or antelopes, ash trees or maple trees? Are we wrongly "playing God" when we intervene in nature for the sake of some particular animal or plant? If "everything is connected to everything else" in nature, as modern ecology teaches, what risks might we run by messing around with what Leopold called the individual "cog[s] and wheel[s]" of an intricately interconnected whole? We have a pretty good idea how to promote the long-term good of the Amazon rainforest (mainly by leaving it alone), but do we know what is best for an individual frog or flowering tree that lives there? Often, we create an ethical and ecological mare's nest when we lose sight of the forest for the individual trees.[13]

As we saw in Chapter 6, there are three major objections to ecocentrism: (1) That only individual organisms—and not collective entities such as ecosystems or species—can have moral status or considerability;[14] (2) to sacrifice individuals for the good of ecological wholes is an indefensible form of "environmental fascism";[15] and (3) the kind of holistic thinking favored by ecocentrists—for example, focusing, as Leopold urges, on what must be done

10. Holmes Rolston III, *A New Environmental Ethics: The Next Millennium for Life on Earth* (New York: Routledge, 2012), p. 135.

11. Rolston, *A New Environmental Ethics*, p. 167.

12. Rolston, *A New Environmental Ethics*, p. 167.

13. As we saw in Chapter 5, an added attraction of ecological holism is that it fits with the way most environmental scientists, policymakers, and conservation officers view nature and humanity's proper role within it. A disconnect between environmental ethics and environmental science and policymaking would not be good for the planet.

14. See, for example, Peter Singer, *Practical Ethics*, 2nd ed. (New York: Cambridge University Press, 1993), pp. 280–84; and Harley Cahen, "Against the Moral Considerability of Ecosystems," *Environmental Ethics* 10:3 (1988), pp. 195–216.

15. See, for example, Regan, *The Case for Animal Rights*, pp. 361–62; and Elliott Sober, "Philosophical Problems for Environmentalists," in Bryan Norton, ed., *The Preservation*

to "preserve the integrity, stability, and beauty of the biotic community"[16]—is misanthropic because it would require a huge reduction of human population or other extreme sacrifices to human well-being.[17] Let's briefly consider these three objections.

First, can a collective entity, such as a species of elk or an alpine ecosystem, have interests or other qualities that confer, or contribute to, moral standing?

It is true that neither a species nor an ecosystem can feel pain, joy, or have mental experiences. They have no mind or consciousness.[18] But it does not follow that such ecological wholes cannot have interests. As Aristotle stated, "the whole is something besides the parts."[19] What is good for the Yankees may not be what is good for its individual owners, managers, and players. As Lawrence Johnson argues, it often makes perfect sense to attribute interests to species and to distinguish the questions "What is best for species X? and "What is best for individual member A of species X?"[20] Just as individual organisms can be sick or healthy, thrive or fail to thrive, so too can whole species. In other words, just like individuals, species can be better off or worse off, and thus have well-being interests, and such interests are sufficient for moral status.

When we ask whether ecosystems, as well as species, can have morally significant interests, things become murkier. As we saw in Chapter 5, it is difficult to define "ecosystem" with precision. We may agree that the Florida Everglades are an ecosystem, but what about a puddle of standing rainwater on a decaying stump? Or a small rock that is host to countless microorganisms? Is the Arctic a single ecosystem or many? If many, how many, and how can we individuate them? Setting aside such troublesome conceptual issues, it is puzzling to see how ecosystems, which are collective entities made up of both living and nonliving components, can be objects of direct moral concern. Aren't we inviting confusion when we speak of the interests or well-being of, say, a mountain or

---

of Species (Princeton, NJ: Princeton University Press, 1986); reprinted in Schmidtz and Willott, eds., *Environmental Ethics*, p. 133.

16. Leopold, *A Sand County Almanac*, p. 262.

17. William Aiken, "Ethical Issues in Agriculture," in Tom Regan, ed., *Earthbound: Introductory Essays in Environmental Ethics* (Prospect Heights, IL: Waveland Press, 1984), pp. 269–70.

18. Assuming, of course, that animists and pan-psychics are wrong in thinking that everything in nature is "alive" and ensouled. If as Wordsworth believed, "every flower enjoys the air it breathes," then it becomes easy to see how natural objects can have moral status.

19. Aristotle, *Metaphysics*, 1045a10 (Ross translation), in Richard McKeon, ed., *The Basic Works of Aristotle* (New York: Random House, 1941), p. 818.

20. Johnson, *A Morally Deep World*, pp. 208–9.

a river, as something distinct from, and not fully reducible to, the interests or well-being of the living things that call such places home?

There is, in fact, a clear sense in which ecosystems can and should be objects of direct moral concern. Like an individual plant or animal, an ecosystem can be healthy or unhealthy, biotically rich or biotically impoverished. They can be harmed or benefited. As Leopold notes, a mountain ecosystem is harmed when apex predators like grizzlies and wolves are exterminated, resulting in erosion, loss of biodiversity, and eventually mass starvation due to over-browsing by deer and other animals.[21] We "think like a mountain," as Leopold urged us to do, when we think holistically and recognize that ecosystems, like individual organisms, have intrinsic value and welfare-interests—and hence moral standing.

A second common objection to ecocentrism is that it leads to **environmental fascism**, that is, to an unjustifiable subordination of the individual to the allegedly larger or overriding good of the State (or of some other collective). Ecocentrism, by prioritizing ecological collectives like species and ecosystems over individual organisms, seems to reflect a similar group-trumps-individual mindset. According to critics, it is thus morally indefensible for the same reason that fascism is morally indefensible in human affairs. Is this criticism sound?

As I suggested in our discussion of Aldo Leopold's land ethic, there is a real basis for concern here. In human ethics and in most human legal systems, we generally do not accept any strong form of holism. Most of us believe (or at least affirm) that individuals have "inalienable" rights and an inherent dignity that must be respected, even at great cost to the collective happiness or public welfare. Why should environmental ethics be any different? Why is "fascism" acceptable for deer and rabbits but not for human beings?

I believe that some forms of ecocentrism are vulnerable to this objection. Anyone who seriously believes, for example, that Leopold's summary moral maxim ("A thing is right when it tends to preserve the integrity, stability, and beauty of the biotic community") is the ultimate touchstone of environmental ethics must sometimes be prepared to accept massive suffering and death to individual organisms to achieve the required collective good. Should thousands of rare tropical birds be killed because they have been blown out to an island in a storm and now threaten a common species of gnat that contributes, in a small way, to the integrity and stability of the island's biotic community? To do so would seem to be holism run amuck.

21. Leopold, *A Sand County Almanac*, pp. 139—41.

Fortunately, the brand of ecocentrism I endorse—moderate ecocentrism—isn't vulnerable to this objection. It embraces sensible holism, not eco-fascism. How so? By imposing curbs, or side constraints, on common ways that holism can be pushed too far. First, moderate ecocentrism does not claim that the good of ecological wholes *always* trumps the good of individual organisms. This is a defeasible truth, and there are exceptions.[22] On any defensible version of ecocentrism, a balance must be struck between the morally relevant interests of individuals and those of wholes. Second, moderate ecocentrism does not claim that all organisms are equal in intrinsic value or moral standing. It permits rankings (dogs over dandelions, leopards over lichens, jaguars over jalapenos). It thus would not permit the killing of large numbers of rare birds to prevent the extinction, on a single isolated island, of a common species of gnat. Finally, moderate ecocentrism permits a certain degree of favoritism of human interests over those of nonhuman organisms or ecological wholes. This is another way in which moderate ecocentrism is not purely holistic because it would ordinarily permit, for example, the bulldozing of a meadow to build a badly needed hospital or school. Moderate ecocentrism thus has plausible ways of responding to the charge of eco-fascism that may not be available to other forms of ecocentrism.[23]

The final common objection to ecocentrism is that it is misanthropic—that is, hostile to, or biased against, humans. Specifically, it is claimed, if the prime environmental directive is to promote the health and well-being of ecological wholes, then huge sacrifices of human happiness, prosperity, and well-being would often be required to achieve this. Why? Because if we *truly* wanted to do what is best for nature and truly behave as "plain citizens" of earth's biotic community, we should drastically reduce our population, blow up our dams, abandon most of our cities and our croplands, allow vast swaths of nature to return to wilderness, and move what's left of the human race into small, solar-powered belts of coastal cities, where we would subsist on a simple—perhaps mostly microbial—diet.[24] If this is the practical upshot of ecocentrism, then it is a

22. A view that the most distinguished living ecocentrist, J. Baird Callicott, also defends. See Callicott's *Beyond the Land Ethic: More Essays in Environmental Philosophy* (Albany, NY: State University of New York Press, 1999), p. 73.

23. Of course, many radical environmental ethicists would say moderate ecocentrism is *too* moderate, that it strays too far from the ecocentric straight and narrow and is unduly human-centered. I believe that a moderately human-centered theory is more likely to garner support among environmentalists, as well as being grounded in real differences that set humans apart from nonhuman organisms.

24. A claim made by some of the more radical deep ecologists and their intellectual forbears. For examples, see Bill Devall and George Sessions, *Deep Ecology: Living as if*

wildly impracticable theory that can safely be put on the shelf along with other museum pieces of 1960s nostalgia.

There are extreme versions of ecocentrism that fall victim to this objection. But there are also more sensible forms of ecocentrism, such as Leopold's land ethic, that pretty clearly do not.[25] At any rate, my version of ecocentrism—moderate ecocentrism—is not misanthropic. It requires humans to respect nature, to value nature for its own sake, to tread more lightly on the earth, and to make far greater efforts than we do now to live in sustainable harmony with nature. But it does not require the extreme sacrifices of human interests that would open it to the charge of misanthropy.

In sum, moderate ecocentrism includes most of the things environmentalists like about ecocentrism while also steering clear of serious objections that more radical forms of ecocentrism confront. It fits well with widely accepted environmental values and achieves a reasonable accommodation between competing human and nonhuman interests. For those reasons, I believe moderate ecocentrism (or some close variant of it) is the best and most promising theory of environmental ethics.[26]

Chapter Summary

1. What, in broad strokes, would a satisfactory environmental ethic look like? This chapter sketches and defends a form of moderate ecocentrism.
2. Moderate ecocentrism of the sort I defend consists of four general principles:
    - All living things have inherent worth and deserve moral respect and consideration.

Nature Mattered (Salt Lake City, UT: Gibbs Smith, 1985), pp. 172–76.

25. As noted in the previous chapter, it's easy to be misled by some of the more colorful rhetoric in Leopold's *Sand County Almanac*. A careful reading of his writings as a whole makes clear that he was far less radical than he is often taken to be. Leopold was, in fact, an anthropocentrist by most definitions of that term.

26. Of course, what I have presented is more of a sketch than a theory. A fully worked out theory would need to be far more comprehensive and detailed. In particular, it would need to spell out much more clearly than I have here, how conflicts between human and nonhuman interests should be resolved. For an instructive—though problematic—model of how such priority principles might be formulated, see Taylor, *Respect for Nature*, pp. 256–307.

- Some living things have greater inherent worth than others. Human beings, in particular, have a special dignity and inherent worth. As a result, human interests should generally take precedence over those of other life-forms.
- Some ecological wholes, such as ecosystems and species, have inherent worth and deserve moral respect and consideration.
- In environmental matters, our primary concern should generally be for the health and well-being of ecological wholes, such as species and ecosystems, rather than for the good of individual organisms.

3. All living things have inherent worth and moral status, in part because they have interests and a good of their own. This is true even of organisms harmful to humans such as mosquitoes.
4. It is not the case that all living things have equal inherent worth or equal moral status. There are many value-conferring qualities that contribute to inherent worth in nature, including beauty, sentience, generativity, rationality, and a capacity for emotional response. A dog is a higher and more advanced organism than a dandelion and possesses greater inherent worth because it possesses a greater range (both quantitatively and qualitatively) of intrinsic-value-conferring capacities. Human beings have a special degree of inherent worth and dignity in virtue of their higher-order capacities for rational reflection, autonomous action, rich emotional response, moral discernment, and other advanced capabilities. As a result, humans may, as a rule, permissibly prioritize their own interests over those of plants and nonhuman animals—though all biotic interests must be given their just weight.
5. Some collective entities such as species and ecosystems also have moral status. Like individual organisms, species and ecosystems have welfare interests and can be harmed or benefited. Consequently, they deserve ethical consideration and respect.
6. In environmental decision-making, we should normally be more concerned with the good of species and ecosystems, rather than with the good of individual plants and animals. This is because species and ecosystems typically have greater natural value than individual organisms do, and because—as experience shows—humans have a poor track record when they attempt to "play God" with nature.

Discussion Questions

1. What is moderate ecocentrism? How does it differ from other forms of ecocentrism, such as Leopold's land ethic?
2. Do all living things have inherent worth? Do some organisms have greater inherent worth than others? If so, why?
3. Does all value in nature depend upon human valuers?
4. Do humans have a special and higher-order kind of dignity or inherent worth? Why or why not?
5. Why does the author believe that our primary environmental concern should be with the health and well-being of species and ecosystems, rather than with the good of individual plants and animals? Do you agree?
6. Why does the author believe that it is usually—but not always—best to leave nature alone? Do you agree? If so, can you think of cases where human interference in nature backfired or otherwise had bad effects? Conversely, can you think of cases where it was *good* that humans interfered with nature?

Further Reading

My thinking on moderate ecocentrism is much indebted to the work of two distinguished environmental ethicists: Holmes Rolston III and Lawrence E. Johnson. (Which isn't to say they would agree with everything I say; far from it.) For Rolston, see especially his *A New Environmental Ethics: The Next Millennium for Life on Earth* (New York: Routledge, 2012); and *Environmental Ethics: Duties to and Values in the Natural World* (Philadelphia, PA: Temple University Press, 1988). For Johnson, see his *A Morally Deep World: An Essay on Moral Significance and Environmental Ethics* (New York: Cambridge University Press, 1991).

Part Two

Current Issues

Chapter 8

Environmental Responsibilities to Future Generations

All thoughtful, caring people give some thought to the future consequences of their actions. To think only of immediate, short-term pleasures is both improvident and immoral. Only fools sacrifice long-term, high-value goals (e.g., a college degree, a happy family life, physical health, or a comfortable retirement) for the sake of short-term gratifications. And only ethical nihilists or sociopaths are oblivious to the physical or psychological wreckage their selfish acts may inflict on other lives.

The notion that we should care about the future is also a basic assumption of modern environmental thought. Key environmental notions such as sustainability, resource conservation, wilderness preservation, wildlife management, environmental stewardship, climate-change mitigation, recycling, and protection of endangered species, all presuppose that people alive today should care about posterity and the long-term ecological health of our planet. Few would deny "that we owe it to future generations to pass on a world that is not a used-up garbage heap."[1] In short, the idea that we have ethical responsibilities to future persons and generations is fundamental to both commonsense and to contemporary environmentalism.

Here as elsewhere, however, pesky philosophers insist on raising questions about "commonsense" assumptions. Surprisingly knotty puzzles crop up once we begin thinking seriously about such questions as:

- Do we have ethical *duties* to future generations? If so, what duties? Do such duties include certain environmental obligations? If so, what environmental obligations?
- Do future generations have moral *rights*? If so, do these include rights to certain environmental goods and resources? If so, what environmental goods and resources? And in what amounts?

1. Joel Feinberg, *Rights, Justice, and the Bounds of Liberty: Essays in Social Philosophy* (Princeton, NJ: Princeton University Press, 1980), pp. 180–81.

Let's see if we can get a better handle on these questions. In this chapter, we'll examine some common reasons for denying that we have any environmental duties to future generations. After careful analysis, we'll see that none of those arguments are sound, and that people living today do have important environmental responsibilities to unborn generations.

8.1 Common Reasons for Denying that We Have Responsibilities to Future Generations

We can begin by setting aside two fairly common, but confused, arguments for rejecting ethical duties or responsibilities to future generations.[2]

Some thinkers—ethical egoists—have claimed that we have no ethical duties to future generations, because *we have no ethical duties (or at least no direct ethical duties) to other people at all.* This is an ethical theory we examined in Chapter 1. According to most ethical egoists, humans are naturally and incurably selfish, and are motivated—consciously or unconsciously—exclusively by considerations of self-interest.[3] On this view, since humans *can* care only about

2. The term "future generations" is ambiguous. Does it mean (a) all human beings who will ever exist from the present moment forward (thus including nearly all currently living people); (b) all human beings who will exist after everyone who is now alive is dead (thus excluding all currently living persons and including only fairly remote descendants); or (c) all as-yet-unborn human beings (thus excluding all currently living persons, but including more than just remote posterity)? The second option, (b), is clearly implausible. It would exclude people who are alive in, say, fifty years, since many people now alive will (presumably) still be living then. Options (a) and (c) are also problematic. Option (a) implies, for example, that everyone who will still be alive in two years is part of "the future generation." It thus trivializes the question, "Do we have duties to future generations?" because it is uncontroversial that we have ethical obligations to loved ones and other persons who will still be living a couple of years from now. Option (c) implies, somewhat oddly, that many people who will be alive in, say, thirty years, are not members of a future generation because they are not among those who are yet to be born. Thus, in asking what, if any, duties we owe to the people who will be alive thirty years from now, we must restrict our attention to the younger members of that population—those members who haven't yet been born. All things considered, (a) seems the most defensible definition of "future generation," but with the proviso that our main ethical focus is on the welfare of yet-to-be-born persons and more or less remote descendants. Even more complications arise, of course, if the term "future generations" is expanded to include both human and nonhuman forms of life.

3. Not all ethical egoists are "psychological egoists" in this sense. That is, not all believe that all human actions are ultimately and unavoidably motivated by self-interest. But the belief that people are incurably self-interested is probably the most common reason for the belief that people ought to be self-interested. (If you *can't* be unselfish, it

themselves, it makes no sense to say that they *ought* also to care about other people; instead, each person can, and ought, to seek only his or her own greatest long-term personal good. It immediately follows from this view that *we have no ethical duties or responsibilities whatsoever* to future generations whenever this conflicts with personal self-interest.[4] On the contrary, each person is free to use and abuse the environment in any way he pleases so long as this works to his personal advantage.[5]

Few words, I trust, need be wasted refuting such an extreme and cynical view. As the medievals liked to say, "*ab esse, ad posse*" ("What is actual must be possible"). Obviously, humans *can* care about people other than themselves, because many people plainly do.[6] And just as obviously, anyone who seriously believes that only *his* interests matter, that only *his* good has ethical weight, would be a contemptible egotist and a moral cretin.[7]

Another common but deeply problematic reason for denying that we have ethical obligations to future generations is that there won't *be* any future generations, or at least not very many of them. Such a view is surprisingly common. According to one recent poll, more than four in ten adult Americans believe that we are now living in the "end times" described in the Bible.[8] Though Christians differ on what exactly will occur in the end times, all agree that when the last trump sounds[9] human and earthly history as we've known it will come to an end. Many Christians also believe that in the final destruction of the

---

makes little sense to say that you *should* be unselfish.) For a classic refutation of psychological egoism, see Joel Feinberg, "Psychological Egoism," in Joel Feinberg and Russ Shafer-Landau, eds., *Reason and Responsibility: Readings in Some Basic Problems of Philosophy*, 12th ed. (Belmont, CA: Wadsworth, 2005), pp. 476–88.

4. I speak here of a crude form of ethical egoism. There are versions that are less extreme and less vulnerable to the objections I note.

5. Needless to say, personal self-interest and environmental concern may sometimes overlap. If I am a farmer, for example, it is not in my long-term self-interest to permit my croplands to become infertile or eroded. But as Aldo Leopold powerfully reminds us, self-interest alone is far from an adequate basis for conservation and sound ecological practice. See Aldo Leopold, *A Sand County Almanac with Essays on Conservation from Round River* (New York: Ballantine Books, 1970), pp. 246–51.

6. As Immanuel Kant noted, "human nature is such that it cannot be indifferent even to the most remote epoch which may eventually affect our species, so long as this epoch may be expected with certainty." Quoted in Annette Baier, "For the Sake of Future Generations," in Tom Regan, ed., *Earthbound: Introductory Essays in Environmental Ethics* (Prospect Heights, IL: Waveland Press, 1984), p. 243n2.

7. For a fuller treatment of ethical egoism, see James Rachels and Stuart Rachels, *The Elements of Moral Philosophy*, 5th ed. (New York: McGraw-Hill, 2005), pp. 68–88.

8. Cheryl K. Chumley, "4 in 10 American Adults: We're Living in the End Times," *Washington Times*, September 12, 2013. Web. 12 January 2020.

9. 1 Cor. 15:52.

world (the Apocalypse), all plant and animal life on earth will be permanently destroyed.[10] Such views clearly raise problems for environmental activism and education. Obviously, it is difficult to get people to care about things like recycling, or wilderness preservation, or species extinction, or climate change when they believe that God is going to reduce the earth to ashes in the very near future. But should we accept such a dire prediction?

Assume, for argument's sake, that God exists and that every word in the Bible is wholly true and inerrant. Does it follow that Bible believers should have little concern for the environment or for future generations of humans? Not at all. For of course, such believers may be mistaken in their reading of the Bible or in their interpretation of historical events, as countless previous believers in end-times delusions have been in the past.[11] Most mainline churches and reputable biblical scholars teach that we *cannot* know with confidence when the world will end, and that we must continue, therefore, to be good stewards of God's creation.[12] If a sizable percentage of the world's population cares little for the environment because they mistakenly believe that the Apocalypse is at hand, this could have catastrophic consequences for the future of the planet. Given the enormous stakes involved, one can only repeat to ardent end-times believers Oliver Cromwell's agonized words to the General Assembly of the Church of Scotland (August 3, 1650): "I beseech you, in the bowels of Christ, think it possible that you may be mistaken."[13]

Setting aside these two popular, but unconvincing, reasons for denying environmental responsibilities to future generations, let's examine some better arguments. Three arguments, in particular, have been the focus of extensive debate in recent years. These are:

- The nonexistence argument
- The ignorance argument
- The different-people argument

Let's examine each of these in turn.

---

10. See, for example, Saint Thomas Aquinas, *On the Truth of the Catholic Faith: Summa Contra Gentiles*, vol. 4, translated by Charles J. O'Neill (Garden City, NY: Hanover House, 1957), pp. 348–49.

11. For a fascinating history of end-times delusions in the Christian tradition, see John M. Court, *Approaching the Apocalypse: A Short History of Christian Millenarianism* (London: I. B. Tauris, 2008).

12. See, generally, Roger S. Gottlieb, *A Greener Faith: Religious Environmentalism and Our Planet's Future* (New York: Oxford University Press, 2009).

13. Quoted in Will and Ariel Durant, *The Age of Louis XIV* (New York: Simon & Schuster, 1963), p. 188.

The Nonexistence Argument

The nonexistence argument is fairly simple and straightforward. It claims that future generations have no moral rights, or any other morally relevant properties, because *by definition* future generations *do not yet exist*. On this view, something can have properties or qualities (e.g., *being green* or *being rectangular*) only if it exists. A human being, for example, can have properties like being tall or thin only if that person actually exists. If a thing doesn't exist—if it is literally *nothing*—then it cannot be red, or square, or heavy, or anything else. Properties, on this view, must inhere in something—some kind of thing, or object, or substance. They cannot, so to speak, be free-floating abstract entities. Thus, future generations do not have rights (e.g., a right to clean air or to clean drinking water) because rights are properties, and future generations have no properties, because only existent things have properties, and future generations do not yet exist. Stated more formally, the nonexistence argument runs as follows:

1. Non-existent things, by definition, do not exist.
2. A thing has properties (i.e., qualities or characteristics) only if it exists.
3. Moral rights are properties.
4. Future generations, by definition, do not yet exist.
5. So, future generations have no moral rights.
6. If future generations have no rights, then the present generation has no ethical duties or responsibilities to future generations.
7. Therefore, the present generation has no ethical duties or responsibilities to future generations.[14]

Is this argument sound? Does it succeed in showing that future generations have no moral rights (or other markers or bases of moral considerability) that present generations are bound to respect?

It doesn't; the argument, in fact, is fallacious. It rests on a false picture of how moral duties or responsibilities are grounded.

One way that a moral duty can be generated is by being linked to a corresponding right. For example, if I have a moral right not to be robbed, then you (and everybody else) have a correlative moral duty not to rob me. Rights and

14. For a statement and defense of the nonexistence argument, see Richard De George, "The Environment, Rights, and Future Generations," in Kenneth E. Goodpaster and Kenneth M. Sayre, eds. *Ethics and Problems of the 21st Century* (Notre Dame, IN: University of Notre Dame Press, 1979); reprinted in Ernest Partridge, ed., *Responsibilities to Future Generations* (Buffalo, NY: Prometheus Books, 1981), pp. 157–66.

duties are frequently correlated in this way—a kind of gravitational "ethical pull" (a moral right, such as a right to religious freedom) produces a corresponding "ethical push" (a duty to respect the right).[15] But it is a mistake to think that for every moral duty there is a corresponding right. If I am filthy rich, for example, I arguably have a moral duty to give some of my money to worthy charitable causes. But no particular charity—for example, the Red Cross—has a right to any determinate share of my money. Similarly, if three young children are drowning near a capsized raft, and I can easily and without risk save one—but only one—then I have a moral duty to save one child, even though none of the children have a right to be the one that is rescued. These are examples of what the philosopher Immanuel Kant—not very aptly—calls "imperfect duties" ("imperfect" because there is more than one way to fulfill them and we have some flexibility in how we choose to do so). Such examples make clear that rights are not invariably correlated with duties, and that duties can be produced by morally relevant considerations ("ethical pulls") other than rights.

Having grasped this crucial point, we can now see where the nonexistence argument goes wrong. It may be true, say, that Generation 200 (the group of not-yet-born persons who will be alive two centuries from now) have no presently existing rights. But as we've just seen, presently existing rights are only one way that moral duties and responsibilities can be generated. There might be other factors that impose ethical constraints on things we might do today that would affect Generation 200. What sorts of factors? There are probably many, but consider just one: the duty not to cause unnecessary suffering.[16]

Imagine if people today decided to completely ignore the welfare of future generations and totally trash the environment for the sake of short-term gain. The result? A booming economy and a wild joy ride for the present generation but "a used-up garbage heap" for those who will come after us. As a result of

15. The useful terms "ethical pull" and "ethical push" are borrowed from Robert Nozick, *Philosophical Explanations* (Cambridge, MA: Harvard University Press, 1981), p. 451.

16. Moral responsibilities to future generations might also be grounded in various environmental virtues, including what Ronald Sandler calls "virtues of environmental stewardship" (e.g., justice and benevolence) and "virtues of environmental sustainability" (e.g., temperance and farsightedness). See Ronald L. Sandler, *Character and Environment: A Virtue-Oriented Approach to Environmental Ethics* (New York: Columbia University Press, 2007), p. 82; cf. Thomas Hill Jr., "Ideals of Human Excellence and Preserving Natural Environments," *Environmental Ethics* 5:3 (1983), pp. 211–24. On such accounts, we have responsibilities to future generations because that's the way environmentally conscientious and virtuous persons think and behave.

our generation's selfishness, suppose that 200 years from now millions will die annually from polluted water, toxic wastes, depleted resources, biodiversity loss, rising seas, super-storms, dying oceans, and runaway global warming. It is true that Generation 200 does not *now* possess a right not to suffer those harms. But this does not imply that people today have no duties or responsibilities with respect to Generation 200. We do have a duty not to inflict unnecessary suffering—a duty that extends into the future and includes that generation. This shows why the argument from nonexistence fails.[17]

The Ignorance Argument

The ignorance argument against duties to future generations is also fairly straightforward. Consider again Generation 200—the now-unborn people who will be alive two centuries from now. According to the argument from ignorance, we know practically nothing about those people—who they will be, how many of them will exist, or what their needs and problems will be. Moreover, we have no way of knowing whether any sacrifices we make today will actually benefit Generation 200. Take climate change, for example. Our generation might make extraordinary sacrifices—pay much higher taxes, build enormously expensive seawalls, pay huge amounts of money to rapidly decarbonize the global economy—which might not benefit Generation 200 at all. Why? Because for all we know they might possess now-undreamt-of technologies (e.g., "carbon vacuums," "atmospheric thermostats," "biodiversity replicators," and "sea-level controllers") that allow them to easily and cheaply prevent and mitigate climate change. This radical uncertainty, this almost total lack of knowledge about what Generation 200 might want or need, makes it meaningless to talk of any ethical "duties" or "responsibilities" we owe to them. Or so the argument from ignorance claims.

This is a better argument than the nonexistence argument. The fact that we know so little about remote generations—including whether there will even *be* any remote generations—renders problematic any talk of duties or responsibilities to them. But the argument does not fully succeed. It overlooks two important things.

First, not all future generations are *remote* future generations. Some will presumably be coming along in the not-too-distant future, and may include our children or grandchildren, or at least their children and grandchildren. We can know with confidence what, in general terms, such near-term generations will

17. For further criticisms of the nonexistence argument, see Lukas Meyer, "Intergenerational Justice," *Stanford Encyclopedia of Philosophy*. Web. 17 January 2020.

want and need.[18] They will be human, with human needs, desires, hopes, and vulnerabilities. They will need clean water, fresh air, fertile soils, green spaces, and healthy oceans. These simple facts alone are enough to ground moral concern for their welfare.

Second, respect for humanity requires that we have some concern for possible generations that may exist even in the remote future. Remoteness in time, considered simply as such, has no ethical relevance. A mad scientist who plants a doomsday device deep in the earth, with a timer set to go off in 10,000 years, is still an ethical fiend. Similarly, today's generation would be seriously at fault if, say, it fails to take adequate steps to safeguard its nuclear wastes, causing tens of thousands of unnecessary deaths five centuries from now. A widely accepted moral principle of risk management, the **precautionary principle**, comes into play with respect to such possible future harms. This holds, in one widely quoted formulation, that "when an activity raises threats of harm to human health or the environment, precautionary measures should be taken even if some cause and effect relationships are not fully established scientifically."[19] Such a principle seems to apply not merely to threats of current harms but also to future ones. Thus, the argument from ignorance does not show that we have no duties to future generations.

The Different-People Argument

Perhaps the most intriguing objection to the notion of duties to future generations is the different-people argument.[20] Defended most forcefully by the late Oxford philosopher Derek Parfit, the argument runs briefly as follows: lots of big things humans do change the course of history. World War II, for example, not only profoundly changed world politics, it dramatically altered global

---

18. Similarly, it would be morally wrong for me to conceal a hanging bowling ball on a tall building, knowing that sometime in the not-too-distant future the rope will break and the bowling ball will fall to the ground. The fact that I don't have a clue who might be killed or injured by the falling ball in no way lessens my moral culpability.

19. Wingspread Statement 1998, quoted in Stephen M. Gardiner, "Ethics and Global Climate Change," *Ethics* 114 (2004), pp. 555–600; reprinted in Stephen M. Gardiner et al., eds., *Climate Ethics: Essential Readings* (New York: Oxford University Press, 2010), p. 13. As we shall see in the next chapter, it's not easy to formulate the precautionary principle in a way that works satisfactorily in all cases. But the general notion that threats of serious (but uncertain) harm warrant precautions is widely accepted—indeed, simple common sense.

20. The argument goes by various names, notably "the contingency argument," the "disappearing beneficiaries argument," and the "non-identity-problem." The term "different-people argument" is my coinage.

demographics. The massive number of deaths, movements of populations, new technologies, etc. produced by the war caused a countless number of children to be born who otherwise would never have been conceived. This obvious fact—that major historical events and large-scale human projects can radically affect who will exist in the future—provides a key premise of the different-people argument. For suppose we make some big environmental policy decision today. Suppose, for example, that the United States decides to delay taking serious steps to fight climate change for at least thirty years. This will, in all likelihood, be extremely bad for the planet. To fix ideas, also suppose that this decision by the United States proves to be a leading cause of massive sea-level rise that swamps dozens of major cities worldwide in the year 2250, causing untold hardship and economic loss. Could these future victims rightly blame the United States? Could they justly claim that they were harmed by failure of the United States to seriously address climate change? The answer is No, according to the different-people argument. For the policy of delay adopted by the United States would have been a big historical event, with massive reverberations for the distant future. As a result, many if not all of the people alive on the planet in the year 2250 *would not even have existed* if the United States had adopted a different policy. How, then, could the United States have harmed them? Are they in any way *worse off* because of America's course of action? Has that act in any way caused them to have a lower quality of life than they otherwise would have enjoyed (to wit, as non-existent entities)? It's hard to see how. Thus, the argument concludes, it is meaningless to talk about duties not to harm future generations, because *it is impossible to harm them* no matter what environmental policies we adopt today.[21]

This is a clever argument that has generated enormous scholarly discussion. Many critical responses have been offered, some quite high-powered and technical. But it is easy to see, without delving into metaphysical and biological arcana, why the argument fails. First, it's simply not true that *all* future persons harmed by present environmental actions would not have existed if different actions had been taken. A child born in a remote corner of New Guinea in, say, 2040, might well have been born regardless of what environmental policies the United States chose to adopt two decades before. If that New Guinea child's village is swamped by rising seas in 2100, he might quite justifiably claim that he was harmed by U.S. action.

---

21. For a probing but somewhat advanced discussion of the different-people argument, see Meyer, "Intergenerational Justice," section 3 and accompanying notes.

Second, the different-people argument applies, at best, to *major* environmental actions and policy decisions. It seems unlikely that minor actions will cause many people to be born who otherwise would not have been.

Finally, we must again bear in mind that there are many sources of moral pull other than moral rights. All people today have a duty not to cause unnecessary suffering—a duty that extends, as we have seen, into the future. The alleged fact that no identifiable person alive in, say, 2300, can claim a right not to have been harmed by U.S. environmental policy in 2020, does not mean that American citizens today have no responsibilities to people who may inhabit a very sick, hothouse earth in the year 2300. The ancient adage, "First, do no harm," carries into the future, for as Shakespeare reminds us, "The evil that men do lives after them."[22] By choosing to delay any serious efforts to combat climate change, Americans in 2020 would have culpably set in motion a causal sequence that, by hypothesis, will cause enormous future suffering. As Joel Feinberg remarks in a similar context, common sense requires no more for an act to count as "harm."[23] Intuitively, this seems to be true even in cases where the affected person wouldn't even have existed if the wrongful act had not been committed.

Thus, even if the different-people argument does cast doubt on talk of the moral "rights" of future generations, it does not show that people alive today have no environmental responsibilities to future persons, or that we cannot wrongfully inflict ecological harms on them. And since none of the other arguments we examined shows that either, we should stick with the commonsense view that we do have environmental duties not just to the living, but also to people (and nonhuman organisms) that will likely inhabit the earth long after we are gone.

Chapter Summary

1. This chapter asks whether we, the present generation, have environmental responsibilities to future generations. Most people would say that

22. William Shakespeare, *Julius Caesar*, Act III, Scene 2.

23. Joel Feinberg, *The Moral Limits of the Criminal Law: Harm to Others* (New York: Oxford University Press, 1984), p. 97. (Feinberg is discussing whether, legally, one can "harm" a person who hasn't even been conceived yet. He argues that one can.)

we do, but philosophers have pointed to puzzles with this seemingly commonsense view.

2. One reason why someone might deny ethical duties to future generations is because they embrace ethical egoism. Ethical egoists believe that humans are incurably self-interested, self-regarding creatures, and that the only ethical duty anyone ever has is to seek his or her own long-term self-interest and advantage. We saw that this is an unduly cynical view of human nature, and that purely egoistic behavior would often be blatantly immoral.
3. Another reason why someone might deny ethical duties to future generations is because they believe, on religious grounds, that the world will soon come to an end. We saw that it is easy to fall prey to end-times delusions, and that most reputable biblical scholars believe that it is impossible to know with any confidence when the world will end.
4. Three stronger arguments against responsibilities to future generations were next considered: the nonexistence argument, the ignorance argument, and the different-people argument.
5. The nonexistence argument claims that future generations cannot have rights, since only existing things can have rights, and future generations do not yet exist. We saw that this argument fails because moral duties need not be grounded in corresponding moral rights. For instance, a duty not to cause unnecessary suffering is sufficient to ground many environmental responsibilities to future generations, even if those generations presently have no rights.
6. The ignorance argument claims that it is impossible to know what sorts of environmental conditions future generations might want or need, or even whether the human race will still exist in the remote future—thus making it meaningless to talk of duties to future generations. We saw that this argument fails for two reasons. First, it's highly likely that there will be humans in the not-too-distant future, and we know that they will need clean air, unpolluted water, and so forth. Second, there is at least a fair chance that humans will continue to exist on earth well into the remote future. If so, it would be wrong not to take reasonable precautions now against causing them grave environmental harms.
7. The different-people argument claims that people in the distant future can't be harmed by our environmental actions today because those people wouldn't even have existed if we had taken different actions. We

saw that this argument fails for at least four reasons. First, the identity of some future-persons probably won't be affected by the environmental actions we take today. Second, at best, the different-people argument applies to major environmental actions, not very minor ones, which often presumably have little ripple effects in the future. Third, common sense suggests that wrongfully causing an innocent future person to suffer is generally sufficient to say I have "harmed" them, regardless of whether I can know that person's identity, and even in cases where my wrongful act may have been necessary for them to have existed at all. Finally, even if it succeeds, the different-people argument doesn't show that we have no duties or responsibilities to future generations. For as we've seen, the mere fact that we should not cause people unnecessary suffering is sufficient to ground ethical concern for future generations.

Discussion Questions

1. What is ethical egoism? Is it correct? How, in general, would a consistent ethical egoist treat the environment?
2. What is the nonexistence argument? Is it sound?
3. What is the ignorance argument? Is it convincing? How much can we know about the problems and needs of remote generations?
4. What is the different-people argument? Is it sound? Why or why not?
5. Assuming that we do have environmental duties to future generations, what are some of the most important ones?

Further Reading

Questions dealing with environmental responsibilities to future generations are questions of intergenerational justice. For a helpful but somewhat advanced discussion on issues of justice between generations, see Lukas Meyer, "Intergenerational Justice," *Stanford Encyclopedia of Philosophy*, online. For a useful anthology on the topic, see Axel Gosseries and Lukas H. Meyer, eds., *Intergenerational Justice* (Oxford, UK: Oxford University Press, 2009). On environmental justice between generations, including possible legal rights for future generations, see Richard P. Hiskes, *The Human Right to a Green Future: Environmental Rights and Intergenerational Justice* (New York: Cambridge University Press, 2009). Good introductory discussions of intergenerational environmental justice can be found in Joseph R. DesJardins, *Environmental Ethics: An Introduction to Environmental Philosophy*, 5th ed. (Boston, MA: Wadsworth, 2013), chapter 4; Robin Attfield, *Environmental Ethics* (Cambridge: Polity Press, 2003), chapter 4; and Andrew Brennan and Y. S. Lo, *Understanding Environmental Philosophy* (Durham, UK: Acumen Publishing, 2010), chapter 2.

Chapter 9

Population and Consumption

9.1 Population

Humans, with astonishing rapidity, have overrun the planet. Twelve thousand years ago the human population was only about 4 million. By the birth of Christ, this number had grown to 190 million. Population growth was slow—only about 0.04 percent per year—from the 10th millennium BCE to 1700. Then, with the Scientific Revolution, the growth rate began to increase dramatically. Human population reached a billion in 1803, and 2 billion in 1928. Largely as a result of vaccinations, a growing food supply, and improved sanitation and medical care, the rate of population growth surged even faster in the second half of the twentieth century. By 1960, global population had reached 3 billion and was growing by about 90 million a year. Population increase began to slow in the mid-1960s, due in part to new advances in birth control and the women's equality movement. Yet even though the rate of population growth has been cut in half since the mid-1960s (from a peak of 2.2 percent in 1962 to just over 1 percent today), global population still continues to grow by about 82 million per year, mostly in developing countries. As I write, in July 2020, there are more than 7.8 billion people on the planet. This is expected to grow to 8 billion in 2024 and 9 billion by 2038. According to United Nations' projections, global population should reach a peak of around 11 billion by 2100, and then slowly begin to decline.[1]

There are up-sides to a densely populated planet. More people means additional consumers and increased economic activity. More young workers means a larger tax base and more money to support aging retirees. And, of course, a larger population means a greater chance that the world will someday be blessed with another Einstein, Joe DiMaggio, or Leonardo da Vinci.

On the other hand, there are many reasons for viewing overpopulation with serious concern. More people means more crowding, longer lines, more

1. I get these figures from Max Roser, Hannah Ritchie, and Esteban Ortiz-Ospina, "World Population Growth," *Our World in Data*. Web. 9 April 2020.

snarled traffic, and all the ills of densely crowded cities. Even more seriously, overpopulation takes a grave toll on the planet and raises the threat of human catastrophe.

Consider, first, the risk of catastrophe. This is an issue that was famously dramatized by the eighteenth-century parson and economist Thomas Malthus (1766–1834). In his classic *Essay on the Principle of Population* (1798), Malthus pointed out that simple math imposes limits on population growth. Without food, people will die. Yet as Malthus noted, population growth can easily outstrip the food supply. Left unchecked, population grows geometrically (2, 4, 8, 16, etc.), whereas food supply at best grows arithmetically (1, 2, 3, 4, etc.). To this supposedly iron law of asymmetrical growth there are only two possible solutions: fewer births or more deaths. Malthus was skeptical that people would ever voluntarily have fewer babies. What he decorously called "the passion between the sexes"[2] is a biological constant and shows few signs of diminishing. From these seemingly inexorable premises, Malthus drew the pessimistic conclusion that the world will always be filled with poor, hungry people—in short, that misery is, and always will be, the human condition.

With hindsight, we can see where Malthus went wrong. He failed to account for the power of human ingenuity. In the 1960s, inexpensive and highly effective methods of artificial birth control became widely available, dramatically slowing the rate of population increase. Moreover, today's highly productive globalized food system—made possible by such things as improved seeds, chemical fertilizers and pesticides, mechanized farm equipment, and improved irrigation and transport—has vastly increased the world's food supply. Despite the fact that there are now seven times more people on the planet than in Malthus's day, the percentage of the global population that is chronically hungry is far, far lower.[3]

This doesn't mean, of course, that we no longer need to worry about food security. Population experts estimate that by 2050 food production must be increased by 60–70 percent in order to feed the additional 1.3 billion people that are then expected to inhabit the planet. That will be a huge challenge, particularly since climate change is predicted to reduce crop yields and nearly all the best agricultural land is already fully utilized. Food security experts also worry about the possibility of massive crop losses due to blights, superbugs,

---

2. Thomas Malthus, *An Essay on the Principle of Population* (Mineola, NY: Dover Books, 2007), p. 4.

3. According to the United Nations, approximately 815 million people (10.7 percent of the global population) were chronically malnourished in 2016. "2018 World Hunger and Poverty Facts and Statistics," *Hunger Notes*. Web. 9 April 2020.

biological warfare, or other unforeseen threats to one or more of the dozen or so species of grains that provide most of the world's calories. Thus, Malthus's basic concerns must still be taken seriously.

Of course, food supply is only one reason to worry about human overpopulation. We also must consider the toll a large and rapidly growing population takes on the environment. Nearly all major environmental problems—pollution, biodiversity loss, climate change, water shortages, deforestation, resource depletion, toxic waste, overflowing landfills, to name just a few—are directly related to population. That is why Pope Francis is only partially right when he says, "To blame population growth instead of extreme and selective consumerism on the part of some, is one way of refusing to face the issues. It is an attempt to legitimize the present model of distribution, where a minority believes that it has the right to consume in a way which can never be universalized, since the planet could not even contain the waste products of such consumption."[4] Regardless of how resources are distributed, with 7.8 billion people on the planet and concentrated mostly in the temperate zones, it is inevitable that there will be water shortages, resource depletion, vast amounts of waste, and serious problems of habitat loss. Though high and unsustainable patterns of consumption greatly worsen environmental problems, population also plays a significant role.

To take but one example, consider how overpopulation directly contributes to biodiversity loss. Currently, one in eight species of plants and animals are threatened with extinction.[5] The leading cause of extinction is habitat loss. To grow crops, raise livestock, and build cities, humans have co-opted vast areas of the planet. Partly as a result, populations of insects, birds, amphibians, and many other species are crashing around the globe. Overpopulation plays a direct role in this tragic and unprecedented process.

Nearly all environmentalists agree that population growth is a serious ecological problem. The hard question is what to do about it. Some critics of overpopulation favor harsh, coercive measures. For example, Paul Ehrlich, author of the bestselling book *The Population Bomb* (1968), proposed taxes on large families, incentives for voluntary sterilization, and elimination of all food aid to countries that refused to get serious about population control. Whatever their merits, such proposals are too radical to be adopted by Western industrialized democracies any time soon. More moderate means are likely to be pursued

4. Pope Francis, *Laudato Si'*, (May 24, 2015), §50. Web. 9 April 2020.

5. Isabelle Gerretsen, "One Million Species Threatened with Extinction Because of Humans," *CNN*, May 7, 2019. Web. 14 May 2020.

first, including generous assistance for family planning and efforts to speed up what population experts call the **demographic transition** in poor, high-fertility countries—most of which are now in Africa. The demographic transition refers to a long-familiar pattern in which fertility rates tend to fall rapidly in countries that modernize; reduce poverty rates; and embrace progressive policies such as gender equality, universal education, and old-age insurance. Such a transition would not only dramatically curb global population growth, it would also reduce poverty, hunger, and gender discrimination in the developing world. To achieve such transitions quickly would require significant capital investment and foreign aid. But since such investments are urgently needed anyway in order to combat climate change, such assistance should be provided as soon as possible.

9.2 Consumption

Population is not the only thing wreaking havoc on the environment. Environmental destruction is also fueled by high levels of consumption, particularly in affluent industrialized nations, such as the United States, Japan, Australia, and members of the European Union. Let's begin with some basic facts about consumption.

In economic terms, consumption is the buying or using of goods and services in the economy, usually in ways that reduce those goods or services, or make them unavailable or less available to others. For example, buying a winter coat or eating an apple are forms of consumption because they involve the purchase or use of things of economic value in a way that reduces or eliminates their availability to others. They are goods that are "consumed"—that is, purchased or used up as part of a stream of economic activity. Major consumption items typically include food and beverages, clothing and footwear, housing, energy, transportation, health care, and education.

The world economy generates about $80 trillion of goods and services each year. The United States has by far the world's largest economy, accounting for more than 20 percent of the total. Just four countries, in fact—the United States, China, Japan, and Germany—account for more than 50 percent of the global economy, and the top ten countries jointly account for 66 percent. However, the most rapid growth in GDP and consumption is now occurring in the developing world, not in the richer countries.

Consumption raises a host of important questions. Why do people consume so much? Are current patterns of consumption sustainable? Are such patterns fair and morally defensible? Should people, particularly in affluent countries, significantly reduce their personal consumption for the sake of a healthier environment and a more equitable sharing of resources? Are there ways to substantially reduce consumption without harming people and badly hurting the economy? Let's consider these questions in turn.

Why Do People Consume So Much?

As Madonna sings, we are living in a material world, and it is a fact of modern life that many people, particularly in rich countries, buy lots of unnecessary stuff, waste a great deal, and live high-consumption lifestyles, often in ways that are ecologically harmful. As Mark Sagoff notes, this is initially puzzling because most of the great historical influencers of Western civilization (e.g., Jesus, Socrates, Plato, Aristotle, Augustine, Aquinas, Thoreau, etc.) roundly condemned materialism and hedonistic excess.[6] Despite this, nearly all of the richest countries now have relatively high per capita consumption rates, with Switzerland, Norway, the United States, Luxembourg, and Australia (in that order) leading the pack among larger nations.[7] This seems to suggest that the biggest factor driving high consumption is simply *ability to consume.* In other words, whenever humans *can* consume lots of goods and services, they tend to do so. All the preachments of religion and philosophy seem to have little impact on this simple truth of human nature.

One country in which one often witnesses rampant consumerism is, of course, the United States. As we saw, the United States ranks high in per capita household consumption but does not lead the world, which perhaps isn't surprising given that nearly a third of Americans live in poverty or near poverty and have relatively little discretionary income to spend. Despite this, America has very high rates of per capita consumption. What might account for this?

Probably a lot of factors come into play. True to its puritan and immigrant roots, America has always been a magnet for people seeking a better life and material success, many of whom have in fact been able to achieve "the American Dream." In addition, giant corporations, which spend vast sums on sophisticated advertising campaigns that encourage high levels of consumer spending, have long dominated American economic life. Also, the United States has many super-rich individuals who, true to the general maxim that people with

6. Mark Sagoff, "Consumption," in Dale Jamieson, ed., *A Companion to Environmental Philosophy* (Malden, MA: Blackwell, 2001), pp. 474–75.

7. World Bank, "Households and NPISH Final Consumption Expenditure Per Capita, 2019." Data.worldbank.org. Web. 14 May 2020.

money tend to buy lots of things, have very high rates of per capita consumption. Moreover, surveys indicate that large numbers of Americans believe—mistakenly—that "money can buy happiness."[8] This provides an obvious incentive to pursue personal wealth and to enjoy the sorts of high-consumption lifestyles that wealth makes possible. Finally, the United States has extremely high levels of income inequality, which often gives rise to "relative deprivation" effects among the poor. Though America's poor may be relatively well-off compared to desperately poor people in developing countries, they often feel less well-off than those they see around them. This, in turn, can impel a strong desire to "Keep up with the Joneses."[9] These are likely some of the factors that explain America's comparatively high rate of per-capita consumption.

Are Current Patterns of Consumption Sustainable?

Ecologists commonly speak of an environmental practice as "unsustainable" when it proceeds at a rate that cannot be continued indefinitely (or for a desired period of time) without depleting natural resources or degrading the environment.[10] In this sense, numerous environmental practices are currently unsustainable. Some unsustainable practices, obviously, are more concerning than others. In evaluating such practices, one must consider, among other things, the risk of harm, the magnitude of harm, the immediacy of harm, the subjects of the harm (e.g., whether the practice adversely affects human as well as nonhuman welfare), and whether adequate substitutes are available once the resource is depleted. For many environmentalists, some of the more worrying unsustainable trends include rising levels of carbon emissions, consumption, soil erosion, rates of biodiversity loss, population growth, water usage, deforestation, and

8. Selena Maranjian, "The Verdict's Out! Money Can Buy (Some) Happiness," *The Motley Fool*, September 1, 2019. Web. 15 May 2020. A 2019 survey discussed in this article found that 73 percent of American adults believe that money "absolutely" can buy happiness or "can to some extent." As the article notes, however, social scientists have found few direct links between income and happiness once basic needs are met.

9. See Ryan T. Howell, "What Causes Materialism in America?" *Psychology Today*, March 23, 2014. Web. 15 May 2020.

10. For a helpful discussion of the numerous meanings of "sustainability," see Bryan G. Norton, *Sustainability: A Philosophy of Adaptive Ecosystem Management* (Chicago: University of Chicago Press, 2005), pp. 304—99. Like Norton, most ecologists think of sustainability not just in terms of impacts on human welfare ("weak sustainability"), but also in terms of effects on plants, animals, and the environment in general ("broad sustainability"). In this broader sense, a practice is roughly sustainable when it proceeds at a rate that satisfies the wants and needs of current generations of humans and nonhuman organisms without compromising the ability of future generations of these organisms to satisfy their wants and needs. Cf. Norton, *Sustainability*, p. 363.

waste disposal (e.g., plastics in the ocean). By one oft-quoted estimate, if everyone on the planet consumed as much as average Americans, four earths would be needed to sustain them.[11] Such consumption patterns are neither physically sustainable nor, as we will see, fair to those they may harm.

Are Current Patterns of Consumption Fair and Morally Defensible?

People are often reluctant to make moral judgments about what they perceive as simply lifestyle choices. Few Americans, for example, see anything unethical about affluent people living in large, energy-inefficient houses; building backyard swimming pools in arid regions; flying frequently on expensive vacations; and driving pricey, gas-guzzling SUVs. But as we will see in Chapter 13, climate change forces us to think of ethics in a new light. We now know that today's carbon-intensive lifestyles will likely cause mass extinctions, cost trillions of dollars, and probably kill or displace millions of people in the not-so-distant future. As a result, we urgently need to confront the ethical dimensions of consumption patterns, however much we might like to ignore them.

In asking whether current patterns of consumption are fair and defensible, we quickly find ourselves embroiled in hot-button debates. Some people believe that, in general, the affluent are entitled to live lives of conspicuous consumption because they have worked hard for their money, saved responsibly, labored productively, and (let's assume) not violated anyone's rights in amassing their wealth. In short, some would argue, they *deserve* their luxurious, carbon-profligate lifestyles. Others, with more progressivist or egalitarian or Green leanings, might take a different view. Obviously, this is not an issue on which people are likely to agree.

Luckily, there is no need to dive into the mosh pit of ideological politics to see whether current patterns of consumption are morally defensible. As the climate emergency makes crystal clear, they are not. "First, do no harm" is perhaps the most basic principle of ethics, accepted by all major religions and all leading systems of ethics. People who live high-consumption, carbon-intensive lifestyles seem to violate this core principle of morality by harming the innocent and contributing more than is fair to catastrophic climate change. What follows from this? We'll tackle the ethical and policy challenges of climate change in greater depth in Chapter 13, but here let's continue to focus on the ethics of consumption. Given that current patterns of consumption are not morally defensible, how, in general, should we respond? One obvious question to ask is:

11. Charlotte McDonald, "How Many Earths Do We Need?" *BBC News*, June 16, 2015. Web. 15 May 2020.

Should People, Particularly in Affluent Countries, Significantly Reduce Their Consumption?

In previous chapters we saw that many radical environmentalists (e.g., deep ecologists, ecocentrists such as Arne Naess, and biocentric egalitarians such as Paul Taylor) believe that humans must dramatically reduce their ecological footprint, and in particular that we must greatly reduce human population and adopt simpler, low-consumption lifestyles. In a similar but more moderate vein, Pope Francis has called for an "ecological conversion" that entails substantially decreased production and consumption in rich countries and adoption of low-consumption lifestyles that accord with traditional Christian values such as unselfishness, temperance, generosity, contemplativeness, and otherworldliness.[12] Such thinkers believe, in short, that people now enjoying high-consumption lifestyles have an ethical duty to greatly reduce their respective ecological footprints and adopt significantly Greener lifestyles. Are they right?

They might be, but it's important to keep in mind certain practical realities. First, it may not always be practicable for someone to live a low-consumption lifestyle. They might, for instance, have commitments that require them travel frequently, or to stay in a particular carbon-intensive job, or to live in a particular area in which living a low-consumption lifestyle isn't a real option (for example, because of high housing costs or safety concerns). Second, in some cases shifting to a low-consumption lifestyle may not be achievable without significant personal sacrifice. If becoming an "ecological convert" entails divorce, loss of a valued job, relocation to an unsafe neighborhood, no overseas vacations, a long bike commute to work, or other major hits to quality of life, we must ask whether morality really does require such big sacrifices. Rather than becoming an eco-martyr, might it often be better to express one's commitment to earth-friendly causes in less dramatic ways, for example by giving generously to Defenders of Wildlife, buying carbon offsets, volunteering to energy-retrofit low-income housing, or working actively for political candidates who support Green causes? As in God's kingdom, there may be "many gifts"[13] that the faithful can contribute.

Finally, of course, we must consider the social and economic dimensions of shifting to low-consumption lifestyles. A prosperous economy, such as America's, can of course continue to hum along with quite a few people (impoverished grad students, retirees on fixed incomes, beach bums, hipster baristas living in West-Coast eco-communes, etc.) living ecologically low-impact

12. Pope Francis, *Laudato Si'*, §222 (available online).

13. Cf. I Cor. 12: 4.

lives. But what would happen if, as some propose, *everyone* now living a high-consumption lifestyle should quickly shift to a much lower-impact one? Here we must ask:

Are There Ways to Substantially Reduce Consumption without Harming People and Badly Hurting the Economy?

In thinking about the desirability of adopting a significantly lower-consumption lifestyle, we must consider not only how such a shift would impact our families and ourselves personally but also how it would impact society. What would happen if, as many radical environmentalists urge, *everyone* in, say, America, who is now living a high-consumption lifestyle should become an "ecological convert" and quickly shift to a much Greener one?

Well, since 68 percent of the American economy is based on consumer spending, the economy would immediately take a huge hit. Unemployment would skyrocket. Businesses would go bankrupt. People would lose their homes. Government spending—after the usual debt-fueled stimulus packages—would be slashed. Tourist destinations would become ghost towns. Crime rates would rise. In short, a severe long-term economic depression would result, with no obvious (i.e., spending-based) ways to pull ourselves out of it.[14]

Some readers might find this an acceptable price to pay for a Greener, more sustainable, and more equitable world, but most, I suspect, will not. It is clear that current consumption patterns are harmful, unjust, and unsustainable, and must be changed. But this has to be done in a planned and phased way, with an ample safety net, and with full recognition of the social, personal, and economic costs involved. Assuming that no game-changing techno-fixes to climate change can quickly be found, such a phased approach might seem a sure recipe for complete climate disaster. But the costs of a crash decarbonization of the global economy and a sudden, massive shift to low-carbon lifestyles would be even more severe, and would deprive governments of both the financial

14. Of course, this doesn't imply that it would be wrong or harmful for a given individual to significantly reduce his or her personal consumption, either permanently or for an extended period of time (e.g., in retirement). My argument is directed against those who urge a sudden, large-scale reduction in levels of consumption, either voluntarily or through government compulsion. The 2020 coronavirus global pandemic provides a painful lesson of the economic havoc that rapid, unplanned drops in consumption can produce, even in countries that provide generous safety nets. While a transition to a Greener, lower-consumption world is an ecological and moral imperative, it must be gradual and planned. The end-goal should be an economy that is earth-friendly, sustainable, prosperous, and just.

resources and quite likely the political support necessary to effectively fight climate change over the long haul. Here, as in many environmental matters, a sensible and pragmatic middle path is best.

Chapter Summary

1. According to many ecologists, population growth is a serious environmental concern. In 1960 there were 3 billion people on the planet; today there are 7.8 billion. Though the rate of population growth has begun to slow, world population is still projected to increase to nearly 11 billion by 2100. In his influential *Essay on the Principle of Population* (1798), Thomas Malthus famously predicted that population growth would always outstrip food supply, thus making it impossible to have both rising standards of living and large, sustained population increases. Malthus's fears proved to be exaggerated, but population growth continues to be a major environmental concern both because of continuing food challenges and the toll overpopulation takes on the planet and its human and nonhuman inhabitants. Pollution, deforestation, waste disposal, biodiversity loss, climate change, resource depletion, lack of clean drinking water, and many other environmental problems, are directly related to population growth.
2. Current strategies for slowing population growth focus mainly on non-coercive tactics such as family planning, education, and efforts to speed up "demographic transitions" in nations with high growth rates. Some critics, however, believe that stronger population-control measures are needed.
3. In addition to overpopulation, high rates of consumption are also of serious ecological concern. This is particularly true in affluent countries such as the United States, Germany, and Japan, which tend to have much higher per capita rates of consumption than do developing nations. High-consumption lifestyles raise serious ethical and ecological concerns, in part because they are a major contributor to the climate crisis.
4. Consumption raises a host of important questions, including these: Why do people consume so much? Are current patterns of consumption

sustainable? Are they fair and morally defensible? Should people, particularly in affluent countries, significantly reduce their consumption for the sake of a healthier environment and a fairer distribution of resources? Are there ways to substantially reduce consumption without harming people and badly hurting the economy? It was noted that many of these questions are more complex than they may seem.

Discussion Questions

1. Why did Malthus believe that human population is always bound to outrun the food supply? Why was he wrong? To what extent are his concerns still valid?
2. How concerned should we be about human population growth? What efforts, if any, should be made to check such growth?
3. Why does Pope Francis believe that the real problem is not population growth but rather excessive and unfair consumption by affluent nations and well-to-do individuals? Do you agree?
4. Why do people, particularly in affluent nations, consume so much?
5. Are current patterns of consumption sustainable?
6. Are current patterns of consumption fair and morally defensible? If not, what should be done to address them?
7. Should people, particularly in wealthy nations, significantly reduce their consumption for the sake of a healthier environment and a more equitable sharing of resources? Are there ways to substantially reduce average rates of consumption without harming people and dramatically shrinking the economy? If so, how could this be done?

Further Reading

For good general overviews of ethical issues surrounding population growth, see Cecelia Herles, "Population," in J. Baird Callicott and Robert Frodeman, eds., *Encyclopedia of Environmental Ethics and Philosophy*, vol. 2 (Farmington Hill, MI: Macmillan Reference, 2009), pp. 165–71; and

Clark Wolf, "Population," in Dale Jamieson, ed., *A Companion to Environmental Philosophy* (Malden, MA: Blackwell, 2001), pp. 362–76. In recent decades, Paul and Anne Ehrlich have been leaders in warning of the dangers of overpopulation. Among their many books on the subject, see *The Population Explosion* (New York: Touchstone Books, 1991). For a quite different view, see Julian L. Simon, *The Ultimate Resource* (Princeton, NJ: Princeton University Press, 1981). Simon argues that human ingenuity will solve the population crisis as well as all problems of resource scarcity. Derek Parfit's classic work, *Reasons and Persons* (Oxford, UK: Clarendon Press, 1984), is a brilliant and seminal source on issues of population ethics. For a good general treatment of the ethics of consumption, see Mark Sagoff, "Consumption," in Dale Jamieson, ed., *A Companion to Environmental Philosophy* (Malden, MA: Blackwell, 2001), pp. 473–85. An older but still useful collection of readings on the ethics of consumption is David A. Crocker and Toby Linden, eds., *The Ethics of Consumption* (Lanham, MD: Rowman & Littlefield, 1997). One of the great classics of American literature—Henry David Thoreau's *Walden*—offers a sustained plea for adoption of a low-consumption, "contemplative" lifestyle. Bill Devall does as well, from a deep ecologist's perspective, in his *Simple in Means, Rich in Ends: Practicing Deep Ecology* (Salt Lake City, UT: Peregrine Smith Books, 1988). From a Roman Catholic perspective, Pope Francis makes a case for reduced consumption by affluent individuals and nations, as well as for a more equitable distribution of resources, in his 2015 encyclical, *Laudato Si'* (available online). On the connection of overconsumption to climate change, see Bill McKibben, "A Special Moment in History: The Challenge of Overpopulation and Overconsumption," *The Atlantic Monthly*, May 1, 1998 (available online).

Chapter 10

Food Ethics

Food is life. It is also culture and tradition. Food brings us together. Food expresses who we are and how much we care. We celebrate with food and give thanks for it. Since humans first evolved on the African savannah, the search for food has been the central challenge of all human striving. Food means life or death, feast or famine, gladness or suffering. For these and other reasons, many important ethical issues center around food.

Some principles of food ethics are clear and uncontroversial. Some, in fact, are so basic that they are matters of law. It is illegal, for instance, to tamper with food or to make false claims about its ingredients. In this chapter we will examine four hotly debated issues of food ethics, namely:

- Is it ethical to eat animals?
- Should we eat local?
- Should we eat genetically modified foods?
- What should we do about world hunger?

Let's begin with the first issue: Should we raise and eat animals for food? Or should we, instead, become vegetarians (people who eat no animals) or even vegans (people who refrain from eating or using all animal products, including milk, cheese, and leather)?

10.1 Is it Ethical to Eat Animals?

In Chapter 3, we considered arguments for and against animal rights. There we saw that a strong case can be made for rethinking traditional anthropocentric attitudes toward animals and abandoning the system of industrial animal agriculture that inflicts great suffering on billions of animals each year. The central thrust of what ethicist Ronald Sandler calls "the argument from animal welfare" can be formally stated as follows:

1. Animal agriculture causes very large amounts of suffering.
2. We ought not cause suffering to others without adequate reason.
3. There is no adequate reason for animal agriculture.
4. Therefore, we ought to abandon animal agriculture and adopt a (largely) non-meat diet.[1]

As Sandler notes, this argument focuses only on animal agriculture and does not prove, even if it is sound, that it is always wrong to eat meat. For example, the argument does not show that it would be wrong to eat a deer that had been killed by lightning. But since nearly all the meat that people in developed countries eat is derived from animal agriculture, the argument would, if successful, justify adoption of a mostly vegetarian diet. Animal agriculture seems to be defensible only if we adopt some strong form of anthropocentrism that regards animal suffering as not mattering very much. Yet, as we saw in previous chapters, no strong form of anthropocentrism can be defended. So, Sandler seems to be on strong ground in claiming that we should abandon animal agriculture and switch to a mainly plant-based diet.

Sandler's argument for adopting a non-meat diet is based on animal welfare. Another common argument focuses on the harm animal agriculture causes to the environment.

Raising livestock for food is bad for the environment in lots of ways. Livestock ranching is a leading cause of tropical rainforest destruction and biodiversity loss. It is also a major contributor to climate change. Worldwide, about 11 percent of greenhouse gas emissions are derived from agriculture, and nearly two thirds of agricultural emissions are related to livestock.[2] Cattle, in particular, produce vast amounts of methane, a potent greenhouse gas.

Animal agriculture also damages the environment in other ways. Overgrazing causes erosion, soil compaction, and desertification. Runoff from feedlots pollutes streams and groundwater. Animal agriculture, particularly cattle raising, uses vast amounts of water. In California, for example, it takes several thousand gallons of water to produce a single pound of beef.[3] Livestock also consume huge amounts of edible grain (mainly corn, barley, and oats). This is a highly

---

1. Ronald L. Sandler, *Food Ethics: The Basics* (New York: Routledge, 2015), p. 75 (slightly adapted).

2. Sandler, *Food Ethics*, p. 87.

3. Tristam Coffin, "The World Food Supply: The Damage Done by Cattle-Raising," *The Washington Spectator* 19:2 (January 13, 1993); reprinted in Paul Pojman, ed., *Food Ethics* (Boston, MA: Wadsworth, 2012), p. 149. Exact estimates vary.

inefficient way to produce food. For instance, ten calories fed to cattle results in only one calorie fed to people. In the United States 70 percent of grain production is fed to livestock. Growing this grain requires huge amounts of water, along with millions of tons of chemical fertilizers and pesticides, which impacts human health and causes all kinds of ecological damage, including biodiversity loss and oxygen-deprived dead zones in places like the Gulf of Mexico and Chesapeake Bay. Feeding grain to animals means less food is available to feed the world's hungry. Antibiotics fed to animals contribute to antibiotic resistance in humans and might lead to the creation of superbugs that could cause terrible pandemics. Eating less meat has been shown to be healthy and would greatly reduce the incidence of heart disease, strokes, diabetes, obesity, and cancer. For all of these reasons, abolishing animal agriculture would not only reduce unnecessary suffering in animals but would also be very good for the planet.

10.2 Should We Eat Local?

For most of human history, people had no choice but to eat locally and seasonally. Medieval Europeans, for example, couldn't eat corn, potatoes, or tomatoes, since those grew only in the then-undiscovered New World. And if a medieval European wanted a fresh apple, he had to wait until apples ripened in the early fall. Now an amazing variety of foods from around the world are available in supermarkets and restaurants year-round. This cornucopia, however, comes at a price. Foods shipped long distances may be loaded with preservatives and may not be as tasty or as nutritious as fresher foods that are produced and consumed locally. In addition, fruits and vegetables that come from developing countries may have been picked by agricultural workers who are treated as virtual serfs of the soil and exposed to dangerous pesticides. Finally, nonlocal food often has a large carbon footprint. Many fruits, vegetables, and other perishables must be shipped by air, which produces lots of greenhouse gas emissions. Airfreight, in fact, produces twenty times more planet-warming emissions than does transporting by ship or rail. Thus, people concerned about the environment and social justice have good reason to care about where their food comes from.

One increasingly popular response to such concerns is to become a **locavore**—that is, a person who eats only or primarily local foods.[4] Few people are complete locavores. Someone who lives in Fargo, North Dakota, for

4. What counts as "local"? Definitions vary, but many locavores try to eat only foods that are grown or produced within a 250-mile radius of where they live.

example, would have to give up rice, chocolate, sugar, coffee, tea, spices, seafood, pecans, and most fruits if they wanted to eat only locally grown or produced foods. They would also probably find it difficult to eat enough fruits and vegetables when those are out of season in the frigid North Dakota winters. Few people take local eating to such extremes. But an emphasis on eating local foods is becoming more common. One indication of this is the recent profusion of farmers markets and so-called community supported agriculture services (where, typically, subscribers pay farmers directly for regular deliveries of fresh food).

We have already noted the two major disadvantages of eating only local foods—lots of healthy and desirable foods aren't grown or produced locally, and many desirable local foods are available only seasonally. Those are real drawbacks. What can be said on the other side of the balance sheet? What are the benefits of eating locally?

Locavores are able to eat lots of ultra-fresh food that may be tastier, healthier, and more nutritious than food that is shipped from far away. They also benefit the local economy by buying from local farmers.[5] In so doing, they may help preserve family farms in their area—farms that might otherwise be forever lost to suburban sprawl. Often, people who buy local foods form personal relationships with those who grow and sell their food. This can strengthen community ties and can be a way of reconnecting with the land and with rural and bygone values. Buying local foods may also help protect the environment by reducing the carbon footprint of one's food purchases.

This last alleged benefit—protecting the environment—is actually not as simple as it may seem. Sometimes local foods create *more* greenhouse gas emissions than foods shipped from far away. For example, rice grown in California requires lots of water and energy, so it's actually Greener for Californians to import rice from Asia.[6] Because agriculture in developed countries is so mechanized and relies so heavily on synthetic fertilizers (which are byproducts of the petroleum industry), local foods may actually be quite energy-intensive. One must also consider how far one needs to travel to buy local foods. Driving just five extra miles to visit a farmers market will put as much carbon dioxide into the atmosphere as shipping seventeen pounds of onions halfway around the

---

5. Some people worry, too, about the risks of having the global food supply largely controlled by giant transnational food companies. There were sporadic food shortages in the United States during the early months of the 2020 Covid-19 coronavirus pandemic. Some see a cautionary tale in this.

6. Peter Singer and Jim Mason, *The Ethics of What We Eat: Why Our Food Choices Matter* (Emmaus, PA: Rodale, 2006), p. 148).

world.[7] Thus, it cannot be assumed that eating local foods is necessarily eco-friendly. Overall, however, the growing local foods movement must be seen as a step in the right direction.

10.3 Should We Eat Genetically Modified Foods?

Today, one of the most hotly debated issues of food ethics is whether **genetically modified foods** should be cultivated and consumed. Genetically modified foods (GMO foods, for short) are foods that have had changes introduced into their DNA through a process of genetic engineering. Genetically modified crops were first planted in the mid-1990s, and today a significant percentage of all the corn, soybeans, and cotton grown in the world are genetically modified. In the United States, over 90 percent of all soybeans, corn, and upland cotton crops are now genetically modified.[8] About 75 percent of processed foods available on American supermarket shelves contain genetically engineered ingredients.[9] Worldwide, the largest producers of genetically modified crops are, in order, the United States, Brazil, Argentina, Canada, and India.[10] In many countries, however, the cultivation of genetically engineered foods is banned, including Russia and most countries in the European Union. A few countries also ban or regulate the importation of such foods. Many countries that permit the cultivation or importation of genetically modified foods require that they be clearly labeled. In 2016, Congress passed a law that will result in mandatory labeling in the United States (effective January 1, 2022).[11] Let's briefly look at the pros and cons of genetically modified foods.

The major benefit of genetically modified crops is increased agricultural yields. Through genetic engineering, crops can be produced that are more disease-resistant, salt-tolerant, and drought-resistant. Engineering plants to produce their own pesticides can reduce crop loss, which saves farmers money and avoids the need for general-use chemical pesticides that kill both harmful and

7. Singer and Mason, *The Ethics of What We Eat*, p. 148.

8. USDA Economic Research Service, "Recent Trends in GE Adoption," August 20, 2019. U.S. Department of Agriculture website. Web. 26 April 2020.

9. Center for Food Safety, "About Genetically Engineered Foods" (online). Web. 26 April 2020. (The Center for Food Safety is a nonprofit advocacy organization opposed to GMOs. Exact estimates vary.)

10. M. Shahbandeh, "Global Genetically Modified Crops by Countries 2018, by Acreage," *Statista*. Web. 27 April 2020.

11. Caitlin Dewey, "Mandatory GMO Labels Are Coming to Your Food," *Washington Post*, May 4, 2018. Web. 27 April 2020.

beneficial insects indiscriminately. Plants can also be engineered to be resistant to herbicides, enabling farmers to control weeds without plowing, thereby preserving topsoil and reducing greenhouse gas emissions through fuel savings and enhanced carbon sequestration in the soil. Given that agricultural demand is expected to increase by 60–120 percent in the next few decades,[12] defenders claim that the gains in agricultural productivity made possible by genetic engineering can help feed a hungry planet.

The downsides of genetically modified crops are more difficult to weigh. Despite the fact that 49 percent of adult Americans believe that genetically modified organisms (GMOs) are unsafe,[13] there is no scientific evidence to support that. Since the mid-1990s, billions of people around the world have eaten GMOs without any apparent adverse health effects. Nor should much stock be placed in apocalyptic predictions that genetic engineering will produce super-weeds and super-pests that are resistant to herbicides and pesticides, or that sterility genes inserted into GMOs could be transferred to wild plants, thus endangering the world's food supply, or that genetic engineering will threaten food security by reducing genetic diversity in crop species. Such claims have been extensively debunked by responsible scientists.[14] Much opposition to GMOs seems to be motivated by overblown fears of a dystopian future marked by globalized corporate control, especially of the world's food supply. There are really only two substantial arguments against genetically engineered crops. One is that it is a form of human hubris, an objectionable example of "playing God" with the natural order. The other is that GMOs violate the precautionary principle, a widely accepted standard of risk management, public health, and environmental decision-making.

The playing-God argument can take several forms. If the claim is that genetic modification is wrong because it creates organisms that are products of human agency, and thus "unnatural," the same could be said about Chihuahuas, Angus cattle, grapefruit (a hybrid mix), and modern wheat (another hybrid mix). The genomic changes introduced into genetically modified crops are in fact quite minor compared to the alterations introduced by centuries of conventional breeding and grafting. A stronger form of the playing-God objection is that it is arrogant for humans to assume that they can fully grasp all the risks associated with GMOs. Might large-scale adoption of GMOs, for

12. Sandler, *Food Ethics*, p. 118.

13. Helen Christofi, "Poll: Americans Divided over Safety of GMO Products," *Courthouse News Service*, November 19, 2018. Web. 27 April 2020.

14. See sources cited in Ronald Bailey, "Dr. Strangelunch: Why We Should Learn to Love Genetically Modified Foods," *Reason*, January 2001. Web. 29 April 2020.

example, have unanticipated harmful impacts on long-term human health, or on insect populations, or on traditional farming in developing countries? As we have seen in previous chapters, humans have a long and troubled history of failing to foresee harmful ecological impacts of their interferences with nature. Isn't it prudent, therefore, to ban GMOs until their potential impact on human welfare and the environment is better understood?

This is a common argument against GMOs and it certainly has some force. Whether it is sound depends on complex issues of risk assessment that can best be judged by scientific experts. An important component of such an assessment would be weighing the risks of *not* embracing GMOs in a world that is rapidly growing hotter and more crowded.

The second major argument against GMOs is similar to the first. It involves the precautionary principle that we briefly considered in Chapter 8, a widely invoked form of the familiar adage that it is "better to be safe than sorry." Versions of the precautionary principle have been included in several international environmental documents, including the World Charter for Nature (1982) and the Rio Declaration (1992). There are many different versions of the precautionary principle. "Strong" formulations of the principle require regulatory actions when there are potential threats to human health or to the environment, whereas "weak" versions merely permit such regulations. The best-known formulation is the so-called Wingspread Statement (1998), which states: "When an activity raises threats of harm to the environment or human health, precautionary measures should be taken even if some cause and effect relationships are not fully established scientifically."[15] Since this mandates that precautions "should" be taken, this is a strong version of the precautionary principle.

No one can quarrel with the commonsense idea that precautions should usually be taken when there are significant risks of harm. Such precautions are commonplace in public health law and environmental management, and indeed in everyday life. But as critics have noted, there are problems with many formulations of the precautionary principle. Consider the Wingspread Statement again. For one thing, it is quite vague. What sorts of "precautionary measures" are called for? Are rigorous testing or regulations enough, or should the activity be completely banned? Moreover, as Cass Sunstein points out, precautions themselves can create risks. Harms can result from both action *and inaction*. For example, thousands of people might die if approval of a new life-saving vaccine is unnecessarily delayed. Thus, some critics charge that the

15. Quoted in Sandler, *Food Ethics*, p. 123.

precautionary principle is self-contradictory. In some contexts it seems to both require and forbid precautions when there are threats of harm.[16]

Finally, the precautionary principle, if read literally, "would be a recommendation for not doing anything of consequence, as all manner of activities 'raise threats of harm to human health or the environment.'"[17] Contrary to the hope and intention of its (mostly progressive) defenders, the precautionary principle thus becomes a recipe for paralysis and preservation of the status quo. Consider whether a new tiger preserve should be established in India to save the last remaining wild Bengal tigers, even though this would require the forced relocation of hundreds of villagers. Though the causal connections aren't scientifically certain, such coerced relocations would certainly raise threats of physical and psychic harm to the villagers. Thus, the precautionary principle seems to imply that no such preserve should be established.[18] This, of course, is not the result most environmentalists would favor; yet it seems to follow from the logic of the precautionary principle.

This reveals a further problem with the precautionary principle. Most of its supporters don't wish it to be interpreted literally. They want it to be construed "reasonably," by which they mean selectively, to support causes they favor. But such a malleable principle is problematic for serious policy analysis. Any defensible version of the precautionary principle—if there is one—should not be ideologically slanted and, in particular, should not be unduly biased against new solutions and technologies. Given the proven benefits of genetically modified foods, it's not clear, in short, that the precautionary principle should bar their production or consumption.

Where does this all leave us? At this juncture, the pros of genetically modified crops would seem to outweigh the cons. There is no evidence that such crops are unsafe and they have clear human and environmental benefits, including increased farm incomes, higher yields, reduced use of herbicides and pesticides, fewer greenhouse gas emissions, and less topsoil loss. Any real risks posed by GMOs should, of course, be rigorously investigated—as a reasonable application of the precautionary principle would suggest. But given human population growth (expected to approach ten billion by 2050) and the

---

16. Cass R. Sunstein, "Throwing Precaution to the Wind: Why the 'Safe' Choice Can Be Dangerous," *Boston Globe*, July 13, 2008. Web. 29 April 2020.

17. Jonathan Adler, "The Problems with Precaution: A Principle without Principle," *The American*, May 25, 2011. Web. 30 April 2020.

18. Or so it seems. But the precautionary principle also requires us to consider harms to the environment. Presumably, extinction of tigers in the wild would be an environmental harm. But how do we balance harms to human health versus harms to the environment? The precautionary principle doesn't tell us.

unprecedented global crisis of climate change, a cautious approval of genetically modified foods seems to make sense.

10.4 What Should We Do about World Hunger?

For all the bad news in the world today, there is much good news that often goes unreported. One terrific piece of ongoing good news is the recent dramatic decline in global poverty. In 1990, 36 percent of the world's population lived in extreme poverty.[19] Today only 10 percent does (mostly in sub-Saharan Africa).[20] That's incredible progress, but it still means that over 700 million people struggle daily with chronic malnutrition and over 10,000 children die of starvation each day. It takes only $15 to feed a hungry child for a month. Do affluent and moderately affluent people have a moral duty to give at least a small amount of their spare income to fight world hunger? Let's consider a classic argument that they do.

In 1971, there was a severe famine in Bangladesh, a poor and densely populated country next to India, caused by a combination of civil war, chronic poverty, and widespread flooding. Millions were displaced from their homes and were starving. Relief efforts were slow and hopelessly inadequate. In response, ethicist Peter Singer published an article in 1972 titled "Famine, Affluence, and Morality"[21] that challenged readers to take personal responsibility for addressing global hunger. Singer's central argument can be summarized as follows:

1. If it is in our power to prevent something very bad from happening, without thereby sacrificing anything else morally significant, we ought, morally, to do it.
2. Severe malnutrition and death by starvation are very bad.
3. People in relatively affluent countries who can afford to do so can prevent some severe malnutrition and death by starvation without sacrificing anything of moral significance by donating a small amount of money to famine relief funds.

---

19. Currently defined by the World Bank as living on less than $1.90 per day.

20. The World Bank, "Poverty," April 16, 2020. Web. 19 July 2020.

21. Peter Singer, "Famine, Affluence, and Morality," *Philosophy and Public Affairs* 1:3 (Spring 1972), pp. 229–43; reprinted in Joel Feinberg and Russ Shafer-Landau, eds., *Reason and Responsibility: Readings in Some Basic Problems of Philosophy*, 12th ed. (Belmont, CA: Wadsworth, 2005), pp. 631–39. References to Singer's article will be to this anthologized edition.

4. Therefore, people in relatively affluent countries who can afford to do so should donate at least a small amount of money to famine relief funds.[22]

Singer thinks that nearly everyone will accept the first premise, but he recognizes that there are some people—sometimes called moral libertarians—who deny that we ever have a moral duty to assist others in need. On their view, any donations to the needy are acts of *charity*, not moral obligation. Why? Because they believe that all moral duties are **negative duties**, that is, duties to refrain from acts that harm other people or violate their rights. They do not recognize any **positive duties**, such as making active efforts to assist others in distress or to donate to good causes.

But is it true that there are no positive duties? More specifically, is it the case that we are never obligated to help a person in distress? Singer offers a powerful example to refute such a view. He writes:

> [I]f I am walking past a shallow pond and see a child drowning in it, I ought to wade in and pull the child out. This will mean getting my clothes muddy, but this is insignificant, while the death of the child would be a very bad thing.[23]

Singer rightly argues that this seems a clear case where there is a positive duty to help a stranger in need. Anyone who refused to save the child because, say, they were late for a hair appointment, would rightly be viewed as a selfish moral reprobate. Singer thus takes it as obvious that if we can prevent something very bad from happening at no real cost to ourselves or without sacrificing anything of moral significance, then we have a moral duty to prevent that very bad thing from happening.

But is this principle really obvious? What if there are *lots of* really bad things I could prevent with little effort or sacrifice, but it's impossible for me to prevent them all? This, in fact, is the situation we face with hunger relief and other forms of humanitarian aid. Severe malnutrition in poor countries is one

---

22. Singer, "Famine, Affluence, and Morality," pp. 632–33. Singer actually endorses a much more contestable claim—that people who can afford to do so should give *generously* to famine relief—so generously, in fact, that if they gave any more, they would be even poorer than the starving and desperately poor people they were trying to help (Singer, p. 634). (In a later work, Singer suggests giving a round 10 percent of one's annual income might be a reasonable goal for most reasonably well-off people. Singer, *Practical Ethics*, 2nd ed. [New York: Cambridge University Press, 1993], p. 246.) Recognizing that few readers would accept such an extreme claim, Singer retreats to the more moderate claim, (3).

23. Singer, "Famine, Affluence, and Morality," p. 633.

serious humanitarian problem in the world. Others include homelessness, lack of clean drinking water, poor sanitation, squalid living conditions, lack of basic medical care, genocide, torture, slavery, persecution, rampant discrimination, and lack of basic educational opportunities. Suppose I can easily afford to give, say, 2 percent of my annual income to humanitarian causes. Singer says I must give at least some of my money to hunger relief—that it would be morally wrong not to do so. But why? Why can't I give all 2 percent to, say, building schools for impoverished girls in Afghanistan, or to providing clean drinking water to villagers in Africa? Why, in a world filled with hurt and woe, must I prioritize one particular good cause—fighting world hunger?[24]

Singer might reply that of all the "very bad" preventable things happening in the world today severe malnutrition is the worst, and that we have a duty to prevent greater evils over lesser ones. But this is dubious for two reasons. First, it's far from clear that severe malnutrition is the greatest preventable evil. Why is chronic hunger worse than dying of some horrible disease, or genocide, or systematic torture? Second, only a utilitarian would claim that we always have a moral duty to prevent greater evils over lesser ones. If I could save five strangers from a runaway trolley by pushing one innocent bystander onto the tracks, should I do so? Most people would say—contrary to what utilitarians would claim—that in this case we should choose the greater evil (five deaths) over the lesser evil (one death).[25] (As we saw in Chapter 1, utilitarianism faces other powerful objections as well.) Thus, Singer has no good reply to the objection that we have no duty to donate to hunger relief organizations, because there are many worthy causes out there that one could give to, and it is implausible to think that hunger relief must be prioritized in the way Singer suggests.

---

24. Another common objection to Singer is that we should first "take care of our own"—the hungry and misfortunate in our own families, communities, and nation—before we think about helping those in distant countries. Singer's response is that there are good utilitarian reasons for giving some "small degree of preference" to one's family, friends, and fellow nationals, but that this slight preference is "decisively outweighed by existing discrepancies in wealth and property." Singer, *Practical Ethics*, p. 234. In other words, starving people overseas are much poorer and worse off than our comparatively privileged family members, friends, and fellow citizens in affluent nations, so we should focus our assistance on the overseas poor to address the greater need. This response is plausible only if one assumes, as Singer does, that an act is morally right only if it maximizes net utility. Giving $100, say, to a pediatric cancer hospital in one's hometown does less good than giving that $100 to starving children in Sudan, so, on Singer's utilitarian view, it would be wrong to give the money to the cancer hospital. Few would agree with this conclusion or, in general, with Singer's utilitarian assumptions.

25. This is a version of the famous Trolley Problem, first introduced by ethicist Philipa Foot in a 1967 paper. For details, see Jack Hacker-Wright, "Philippa Foot," *Stanford Encyclopedia of Philosophy* (online). Web. 24 July 2020.

Might Singer's argument be sound if it were broadened to include not simply hunger relief but all the grave evils of extreme poverty (inadequate shelter, lack of clean drinking water, poor health services, inadequate educational opportunities, etc.)? That's precisely what Singer does in a revised version of his argument. In the second edition of his widely read 1993 book, *Practical Ethics*, Singer reformulates his earlier argument as follows:

1. If we can prevent something bad without sacrificing anything of comparable [moral] significance, we ought to do it.
2. Absolute poverty is bad.
3. There is some absolute poverty we can prevent without sacrificing anything of comparable moral significance.
4. Therefore, we ought to prevent some absolute poverty.[26]

This argument is better in some ways than his earlier argument but worse in others.

It is better because it focuses on more than just world hunger. As we've seen, chronic hunger is certainly very bad, but so are other problems linked to extreme poverty, such as inadequate shelter, poor medical care, and poor sanitation. Recognizing this, Singer came to abandon his claim that we must prioritize hunger relief and now speaks instead of providing overseas aid and fighting "absolute poverty." This is a more convincing view, though many would question Singer's claim that we have a duty to give specifically to aid organizations combating extreme poverty, especially "overseas."[27] If I'm a college student struggling to pay tuition and choose to give all my spare change to help feed a struggling homeless family in my neighborhood, and give nothing to fight poverty or hunger overseas, have I done wrong? It's hard to see why.

That's one problem with Singer's argument. Another is the way he revises the first premise of his original argument.

Singer's new first premise differs from his earlier version in two important respects. First, it requires that we prevent "bad" outcomes, not just those that are "very bad."[28] We'll see how this makes Singer even more vulnerable to one

26. Singer, *Practical Ethics*, pp. 230–31. Singer expands this argument in his later book, *The Life You Can Save: Acting Now to End World Poverty* (New York: Random House, 2009). There, too, Singer argues that our duty is to give to humanitarian aid groups, not necessarily to hunger-relief organizations.

27. Singer, *Practical Ethics*, p. 242.

28. In his original (1972) argument for famine relief, Singer first formulates his key premise as follows: "[I]f it is in our power to prevent something bad from happening, without thereby sacrificing anything of comparable moral importance, we ought, morally, to do it" (Singer, "Famine, Affluence, and Morality," p. 632). For strategic reasons,

frequent objection to his argument. Second, Singer no longer allows us, as he did in his original argument, to avoid giving aid if by so doing we would sacrifice something of "moral significance." Now he permits us to avoid giving humanitarian aid only if by giving aid we would sacrifice something of "comparable moral significance." This is a far more demanding requirement. Few things are of comparable or greater moral significance than saving children from starvation and extreme poverty. Thus, Singer allows fewer excuses, so to speak, for failing to give. Let's reflect on why this weakens Singer's argument.

Let's start with the issue of what Singer considers allowable excuses for not making donations to fight hunger and extreme poverty. Suppose you have a spare $1,000 and you are considering three ways to spend it:

> Option A: Give the $1,000 to overseas hunger relief.
>
> Option B: Give the $1,000 to a reputable international humanitarian organization, to be used in whatever way the organization thinks would do the most good (e.g., providing clean drinking water, administering vaccines, educating poor girls, building a refugee camp, etc.).
>
> Option C: Use $500 to buy a new set of golf clubs for yourself and give $500 to your favorite charity: a pediatric cancer hospital where your younger brother was treated for cancer before he died as a young child.

In his original version of the famine relief argument, Singer argues that you have a moral duty to do A, that is, give the money to overseas hunger relief. As we've seen, this is a mistake, as Singer later recognized. By opting for A rather than B or C, you would be giving up something of "moral significance," namely not giving to other worthy causes. So what really follows from Singer's original argument is that *none* of the three options is morally obligatory. This is a plausible conclusion but not the one Singer wants. On later reflection, Singer came to believe that option C is *not* morally permissible and that one is morally obligated to do *either* A or B (i.e., both options are permissible, but one is duty-bound to do one or the other). His revised argument is designed to produce this outcome. Since Option C is not of "comparable moral significance" to either A or B, C is not an ethically acceptable option. As Singer sees

---

he later retreats to the weaker and less controversial claim that "if it is in our power to prevent something very bad from happening, without thereby sacrificing anything morally significant, we ought, morally, to do it" (Singer, "Famine," p. 633). His official argument, therefore, does not include the stronger claim, though it is clear that Singer personally accepts it.

it, C is something a moral slacker or lightweight would do. Genuinely ethical people—people who truly do their moral duties—are moral saints and heroes who never choose to do things of lesser moral import. They work tirelessly and unselfishly for big (in fact, maximally good) ethical ends.

Now we can see why Singer's revised hunger relief argument, like his original argument, is unconvincing. It sets far too high a moral standard. Singer's argument requires us to be ethical busy bees, constantly working to prevent bad things from happening in the world. We might wish to spend our spare money going to a concert or buying a six-pack of craft beer. But Singer thinks doing either of these things would be morally wrong. Neither is of "comparable moral significance" to feeding a poverty-stricken, hungry child. Nor is the issue only one of how we spend our money. Singer's argument also has radical implications for how we should spend our time. Suppose I have a spare evening and I am considering two options on how to spend it:

Option A: Watch my favorite sports team on TV.

Option B: Volunteer for a few hours at the local soup kitchen.

Since Option A is not of comparable moral significance with Option B, Singer's revised argument *requires* me to opt for B. It demands, in other words, that I be a full-time, tireless ethical do-gooder.

This is far too demanding a moral standard and would leave us with little freedom to live our lives in a way we find enjoyable and worthwhile. As we saw in Chapter 1, a standard objection to utilitarian moral theories is that they require us to be utility-maximizers—that is, utilitarian moral saints. Singer is a utilitarian and his revised hunger relief argument is vulnerable to this objection.[29] In fact, Singer's revised argument is even more vulnerable to this objection than his original argument was. This is because his original argument required us to make personal sacrifices only in order to prevent "very bad" things from happening. In his revised argument, Singer broadens this to include all "bad" things. Since there are more "bad" things in the world than there are "very bad" things, Singer's revised argument imposes an even stricter moral standard. The upshot is that neither of Singer's arguments succeeds in

29. Strictly speaking, Singer's revised hunger relief argument does not presuppose utilitarianism, since it does not require that we actively *do good* but only that we *prevent bad* things from happening whenever we can do so without sacrificing something of comparable moral significance. But like utilitarianism, Singer's argument does require that we be tireless evil-preventers, which, like utilitarianism, sets too high a moral standard.

showing that we have an ethical duty to give to hunger relief or aid groups that fight extreme poverty.

This doesn't mean, of course, that there aren't *other* good arguments for such a duty. For reasons we have seen, utilitarian arguments like Singer's won't work, but other moral theories, such as virtue ethics or pluralistic duty ethics, might do the trick. A virtue ethicist, for example, might argue that virtues such as benevolence, compassion, generosity, and sympathy support a duty to provide humanitarian aid. In like fashion, a pluralistic duty ethicist might appeal to a basic prima facie duty such as beneficence ("Do good") to make a case for humanitarian assistance. Another strategy would be to appeal to standards of ordinary (or "commonsense") morality—that is, to ethical norms that are widely shared in one's society—to argue for a duty of humanitarian assistance. Ethicist Mylan Engel Jr. has offered an interesting argument along these lines. Let's briefly consider Engel's argument.

Engel claims that nearly all of us hold moral beliefs that support the following relatively moderate and undemanding moral principle:

> If you can help to reduce the amount of unnecessary suffering in the world with minimal effort on your part, with no risk to yourself or others, with no noticeable reduction in your standard of living or the standard of living of your dependents, and without thereby failing to fulfill any more pressing obligation, then you ought to do so.[30]

Engel then goes on to argue that this principle supports what he calls a "context-dependent" duty to give to humanitarian causes. Such a duty is context-dependent, he says, because many people can't afford to give, or could do so only at great sacrifice. His argument, therefore, is directed only at those who can easily afford to provide assistance. Such individuals, he argues, have a duty "to provide modest financial support for famine-relief organizations and/or other humanitarian organizations working to reduce the amount of unnecessary pain, suffering, and death in the world."[31]

Engel's argument has two big advantages over Singer's. Unlike Singer's original argument for famine relief, it recognizes that there are many worthy humanitarian causes out there, and there is no duty to donate specifically to hunger relief. More importantly, Engel's argument is not overly strict. It does not assume, as utilitarians do, that we must give and work unstintingly to prevent "bad" (or "very bad") things from occurring. As Engel notes, his argument

---

30. Mylan Engel Jr., "Hunger, Duty, and Ecology: On What We Owe to Starving Humans," in Paul Pojman, ed., *Food Ethics* (Boston, MA: Wadsworth, 2012), p. 137.

31. Engel, "Hunger, Duty, and Ecology," p. 131.

is "minimalistic" and imposes no duties that are not easily met. For that reason it is likely to have wider appeal than Singer's argument does.

The major question I have with Engel's argument is whether it succeeds in showing that most of us have a duty to give specifically to "famine-relief organizations and/or other humanitarian organizations." I might wish to give all my spare income to my favorite environmental group, my college alma mater, or to my local parish. Do these count as "humanitarian organizations"? If not, then Engel's argument seems too limiting. For as we saw with Singer, there are many worthy causes out there that deserve financial support. Even if well-off people do have an obligation to support some good causes, it's not clear that those must include famine relief or even Engel's "humanitarian organizations" that fight "unnecessary pain, suffering, and death." Though as a rule charitable donors should probably prefer high-impact giving to low-impact giving, there is no ethical duty to "do the most good you can do," as Singer claims.[32] Such a view presupposes a utilitarian ethic that we have seen good reason to reject.

A strong case might be made on grounds of compassion or beneficence that well-off people have an ethical duty to give at least some small portion of their income to alleviate famines, extreme poverty, or other major humanitarian disasters—or, in the alternative, to give to other good charitable causes. In other words, a strong argument might be made to support a general duty of charitableness or generosity or compassionate liberality. But such an argument would be nuanced and quite different from Singer's famous hunger-relief argument.

Chapter Summary

1. Many important ethical issues center on food. In this chapter we focused on four issues of food ethics: Should we eat animals for food? Should we eat local? Should we produce and consume genetically modified foods? And what should we do about world hunger?
2. Raising and killing animals for food causes unnecessary suffering, is wasteful of food calories, is generally bad for human health, and causes numerous environmental harms. On these grounds, a strong case can

32. Peter Singer, *The Most Good You Can Do: How Effective Altruism Is Changing Ideas about Living Ethically* (New Haven, CT: Yale University Press, 2015).

be made for switching to a mainly plant-based (that is, vegetarian or vegan) diet.

3. Eating locally grown and produced foods has many benefits. Such foods are generally fresher, more nutritious, and more flavorful than non-local foods; eating local foods usually has a lower carbon footprint; and purchasing locally produced food can strengthen community bonds. Eating *only* locally produced foods is difficult, since some desirable foods may not be grown or harvested locally, and other such foods may be available only seasonally. All in all, though, eating more locally grown food should be seen as a positive trend.
4. Many people today are opposed to producing and consuming genetically modified foods (GMOs). Though such foods have been widely consumed since the mid-1990s and there is no scientific evidence that they are unsafe, they are not permitted to be cultivated in many countries (though nearly all countries allow such foods to be imported). The major benefits of genetically modified foods are higher crop yields, lower costs, reduced pesticide use, less soil erosion, and reduced carbon emissions. Common criticisms of genetically modified crops included concerns about "playing God" with nature, fears about control of the world's food supply by a globalized elite, loss of crop diversity, and possible harms to human health and to nature—risks that critics claim warrant precautions until the effects of GMOs are better known. On balance, the benefits of GMOs seem to outweigh the risks, so a strong (but not slam-dunk) case can be made that they should be allowed.
5. Ethicist Peter Singer has argued that well-off people have an ethical duty to give at least a small amount of money to fight world hunger and extreme poverty. According to Singer, whenever we can prevent something very bad from happening without sacrificing anything of comparable moral significance, we have a duty to do so, and giving to hunger relief satisfies those conditions. Hunger relief, however, is only one of many ways of preventing very bad things from happening at relatively little cost or effort. So, although affluent people may have a duty to give to some worthy causes, they have no duty to give specifically to famine relief (or, more broadly, to aid groups that fight extreme poverty). Moreover, as a utilitarian, Singer assumes that people always have a duty to maximize good consequences over bad consequences. This implies that all who can afford to do so have a duty to give generously

and to work unstintingly to combat global hunger and extreme poverty. Many critics see this as too strict a moral standard.

6. Mylan Engel Jr. has offered a more plausible argument for a duty to give to hunger relief and humanitarian aid. Rather than appealing to utilitarianism or some other controversial moral theory, Engel argues that most of us already hold moral beliefs that support a modest duty to give to hunger relief and/or other humanitarian causes. The major concern with Engel's argument is whether it permits giving to worthy charities of one's choice, rather than to famine-relief or humanitarian aid specifically. If not, it too may be too strict.

Discussion Questions

1. What is Ronald Sandler's argument from animal welfare? Is it sound?
2. Why is animal agriculture generally bad for the environment? Is this a sufficient reason to abolish or greatly reduce animal agriculture?
3. What are the benefits of eating local? What are the drawbacks? In general, is eating local a trend that should be encouraged?
4. Should genetically modified foods be produced and sold? If so, should they be labeled so that consumers know what they are eating?
5. How does Peter Singer argue for a duty to give to hunger relief? How does his revised argument differ from his original argument? Is either argument sound? Do relatively well-off persons have a moral duty to give to hunger relief or to other humanitarian causes? Why or why not?

Further Reading

Good sources on the philosophy of food include David M. Kaplan, *Food Philosophy* (New York: Columbia University Press, 2020); and David M. Kaplan, ed., *The Philosophy of Food* (Berkeley, CA: University of California Press, 2012). For a compact and highly accessible introduction to food ethics, see Ronald L. Sandler's *Food Ethics: The Basics* (New York: Routledge, 2015). For a more in-depth introduction, see Anne Barnhill, Mark Budolfson, and Peter Doggett, eds., *The Oxford Handbook of Food Ethics*

(New York: Oxford University Press, 2018). For two good collections on food ethics, see Gregory E. Pence, ed., *The Ethics of Food: A Reader for the 21st Century* (Lanham, MD: Rowman & Littlefield, 2002); and Paul Pojman, ed., *Food Ethics* (Boston, MA: Wadsworth, 2012). For contrasting perspectives on the ethics of hunger relief, see Peter Singer, *The Life You Can Save: How to Do Your Part to End World Poverty* (New York: Random House, 2009); Onora O'Neill, "The Moral Perplexities of Famine and World Hunger," in Tom Regan, ed., *Matters of Life and Death: New Introductory Essays in Moral Philosophy*, 2nd ed. (New York: McGraw-Hill, 1986), pp. 294–337; and Mylan Engel Jr., "Hunger, Duty, and Ecology: On What We Owe Starving Humans," in Paul Pojman, ed., *Food Ethics* (Boston, MA: Wadsworth, 2012), pp. 129–47.

Chapter 11

Wilderness Preservation

"In Wildness," Thoreau famously declared, "is the preservation of the World."[1] America is blessed with an extensive system of federally designated wilderness areas, amounting to nearly 5 percent of the landmass of the United States. Each year, millions of hikers, backpackers, and other wilderness enthusiasts venture into the American backcountry to experience nature in its mostly primeval, undeveloped state. The idea of setting aside large tracts of scenic, nearly pristine public lands for national parks and protected wilderness areas has been called "America's best idea" by documentary filmmaker Ken Burns.[2] Yet there are many critics who question the whole idea of wilderness preservation, or at least what some call "the received view" of wilderness. In this chapter we'll look at the case for and against wilderness preservation.

11.1 The Concept of Wilderness

We must begin by getting clear on what we mean by **wilderness**, for the word has changed meaning over time and is now used in various ways. According to the Oxford English Dictionary, the term "wilderness" derives from the Old English word *wild-dēor-ness*, meaning "the place of wild deer." In the Bible, the Hebrew and Greek terms that are standardly translated as "wilderness" connote an arid, uninhabited wasteland.[3] In this sense, wilderness was seen

---

1. Henry David Thoreau, "Walking," in *The Selected Works of Thoreau* (Boston, MA: Houghton Mifflin, 1975), p. 672.

2. Dayton Duncan and Ken Burns, *The National Parks: America's Best Idea* (New York: Alfred A. Knopf, 2009). Burns credits Wallace Stegner with inspiring the phrase. See Stegner's article, "The Best Idea We Ever Had: An Overview," *Wilderness* 46 (Spring 1983), p. 4.

3. Roderick Frazier Nash, *Wilderness and the American Mind*, 4th ed. (New Haven, CT: Yale University Press, 2001), p. 3. This older usage is reflected in Samuel Johnson's 1755 *Dictionary*, where wilderness is defined as "a desert; a tract of solitude and savageness."

as a hostile and dangerous place, an abode of evil and desolation—the polar opposite of the paradisal Garden of Eden. This negative Judeo-Christian view of wilderness was long dominant in Western civilization and strongly influenced early European settlers' views of the vast American forests and their "savage" native inhabitants.[4] A more positive view of wilderness emerged with the nineteenth-century Romantic movement, when thinkers such as William Wordsworth, Henry David Thoreau, and John Muir encouraged people to view unspoiled nature as a symbol and embodiment of the divine, and as a soul-refreshing escape from the din, artificialities, and corruptions of modern civilization.[5] In the twentieth century, this Romantic view of nature came to be reflected in both dictionary and legal definitions of "wilderness." Thus, *The Merriam-Webster Dictionary* defines "wilderness" as "a tract or region uncultivated and uninhabited by human beings." Call this, for short, the *dictionary definition* of wilderness. To avoid confusion, this dictionary definition should be contrasted with three other senses of "wilderness," which I will call the *de jure view*, the *de facto view*, and the *received view*.

The de jure view of wilderness is the official federal legal definition of wilderness enshrined in the Wilderness Act of 1964. This landmark piece of legislation, ghost-written by long-time executive director of the Wilderness Society Howard Zahniser, defines wilderness as "an area where the earth and its community of life are untrammeled by man, where man himself is a visitor who does not remain."[6] Following Mark Woods,[7] I call this the de jure ("in law") view because it is limited to substantial tracts of federal land that have been formally designated as part of the nation's National Wilderness Preservation System (NWPS).[8]

4. Nash, *Wilderness and the American Mind*, Chap. 2.

5. Nash, *Wilderness and the American Mind*, Chap. 3.

6. The Wilderness Act of 1964, Public Law 88-577, September 3, 1964. The Act goes on to provide a less poetic and more precise definition of (federally protected) wilderness as "an area of undeveloped Federal land retaining its primeval character and influence, without permanent improvements or human habitation, which is protected and managed so as to preserve its natural conditions and which (1) generally appears to have been affected primarily by the forces of nature, with the imprint of man's work substantially unnoticeable; (2) has outstanding opportunities for solitude or a primitive and unconfined type of recreation; (3) has at least five thousand acres of land or is of sufficient size as to make practicable its preservation and use in an unimpaired condition; and (4) may also contain ecological, geological, or other features of scientific, educational, scenic, or historical value."

7. Mark Woods, *Rethinking Wilderness* (Peterborough, Ontario: Broadview Press, 2017), p. 35.

8. Some states have also officially designated and managed wilderness areas, including Adirondack Park in upstate New York. Thus, there can be state analogues of de jure and de facto wilderness.

Since the Wilderness Act was enacted in 1964, Congress has passed over 170 subsequent wilderness laws, which have collectively expanded the NWPS to more than ten times its original size.[9] Whenever Congress addresses a proposal to add an additional area to the NWPS, it must consider whether the area meets the legal criteria of "wilderness," and if so, whether the area should be added to NWPS. Let's call a tract of federal land that meets the legal definition of "wilderness" but may never have been formally designated as such the de facto ("in fact") definition of wilderness.

Finally, we must distinguish what many anti-wilderness critics call the received view of wilderness. This is closely related to the de facto sense of "wilderness" but goes beyond it in some important ways. This becomes clear when one considers a number of common criticisms of the received view. Val Plumwood, for example, argues that the received view is male-biased because it conceives of wilderness as "virgin" land, waiting to be penetrated by men eager to prove their masculinity in terms of physical toughness.[10] Clearly, any such reading goes far beyond any statutory criteria for wilderness designation. Thus, the received view, in Plumwood's sense, cannot be identified with either the de facto or the de jure definitions of wilderness.

In a similar vcin, J. Baird Callicott criticizes the received view for perpetuating "the pre-Darwinian Western metaphysical dichotomy between 'man' and nature."[11] It does so, Callicott claims, because it defines nature in terms of the absence of human beings. On a Darwinian view, Callicott argues, humans are fully part of nature and hence no human-nature dualism is tenable.

Here again, the received view cannot be equated with the Wilderness Act definition of wilderness, because there is no suggestion in the act that humans are not part of nature. In fact, the act makes no attempt to define "nature" at all. It *does* offer a definition of "wilderness"—a specialized one suited to federal law—but again there is no suggestion of any hard-and-fast dichotomy between humans and wilderness. According to the act, designated wilderness areas are to be "protected and managed" so as to preserve their natural conditions; humans may visit wilderness areas to enjoy "outstanding

9. Woods, *Rethinking Wilderness*, p. 35.

10. Val Plumwood, "Wilderness Skepticism and Wilderness Dualism," in J. Baird Callicott and Michael P. Nelson, eds., *The Great New Wilderness Debate* (Athens, GA: University of Georgia Press, 1998), pp. 659, 662.

11. J. Baird Callicott, "The Wilderness Idea Revisited: The Sustainable Development Alternative"; *The Environmental Professional* 13 (1991), pp. 234-47; reprinted in Callicott and Nelson, eds., *The Great New Wilderness Debate*, p. 348. All citations to this article will be to the Callicott and Nelson reader.

opportunities for solitude or a primitive and unconfined type of recreation"; and any human impacts in wilderness areas must be "substantially unnoticeable." Far from implying that humans and wilderness are mutually exclusive, the act assumes that humans are *necessary* to "protect[] and manage[]" designated wilderness areas. If the received view implies that humans are not part of nature, it must be something different from the view of wilderness enshrined in federal law.[12]

Finally, consider Callicott's further criticism that the received view is unscientific because it wrongly assumes that nature is basically static and that humans must actively manage nature to "preserve" it as it existed at some time in the past, such as prior to European settlement or when the area was formally designated as wilderness.[13] Such a static view of nature is not only inconsistent with modern ecology, Callicott argues, but incoherent, because any attempt to actively manage wilderness to preserve it in some freeze-frame condition would obviously "trammel" it, and thus negate its character as wilderness.

The obvious problem with this criticism is that no serious advocate of wilderness preservation believes that "preservation" means, "keep completely unchanged." Such a view is nonsensical on its face. Does anybody seriously think that, say, the once heavily logged Lye Brook Wilderness in Vermont's southern Green Mountains—just a few miles east of Manchester, Vermont—can or should be restored to anything like the condition it was in in pre-Columbian America, or even as it was when it was added to the NWPS in 1975? Clearly, Callicott's criticism assumes what Michael P. Nelson calls a "purist" notion of

---

12. As John Stuart Mill notes in a classic essay, there are several distinct senses of "nature." One of the meanings Mill distinguishes is "everything which happens . . . without the agency, or without the voluntary and intentional agency, of man." John Stuart Mill, "Nature," in *Collected Works of John Stuart Mill*, vol. 10 (Indianapolis, IN: Liberty Fund, 2006), p. 375. In this sense, "*nature* is all that is not man-made; the *natural* state of anything is its state when not modified by man." C. S. Lewis, *Studies in Words* (Cambridge: Cambridge University Press, 1960), p. 46. This is the sense Callicott apparently has in mind when he equates "nature" with "wilderness" and complains that the received view of wilderness creates an unjustifiable opposition between humans and nature. This criticism is odd, because there are innumerable contexts in which it makes perfect sense to distinguish what is human-made (e.g., art, written speech, a refrigerator) from what is not (e.g., meaningless squiggles in the sand created by the ocean surf). Even the most hardened metaphysical naturalist or materialist would often find it useful to distinguish what-is-human-made from what-results-from-the-spontaneous-forces-of-nature.

13. Callicott, "The Wilderness Idea Revisited," p. 349.

wilderness.[14] According to Nelson, the received view of wilderness is a purist view, because it assumes that true wilderness must be completely uninhabited; totally unmodified by humans; and, if preserved, kept completely unchanged from that time forward. This gives the game away because it makes clear that what Callicott, Nelson, and other anti-wilderness critics call the "received view" is actually a straw man position that no serious advocate of wilderness preservation actually holds.

For all the above reasons, I suggest that we jettison all vague and polemical talk of a received view of wilderness that is different in obviously problematic ways from the widely accepted definition of wilderness long established in federal law. Instead, when speaking of wilderness preservation in the United States, let's stick with the de facto definition of wilderness as public lands that meet the criteria of wilderness under federal law. With this clarification, let's now look at some common arguments for wilderness preservation.

11.2 The Case for Wilderness Preservation

In making a case for wilderness preservation, many of the early preservationists, including John Muir, relied heavily on anthropocentric arguments that stressed human interests and needs. Many of these classic, human-centered arguments still have considerable force. These include:

- *The ecological services argument*: Many wilderness areas provide a host of invaluable ecological benefits. Tropical rainforests, for example, help to stabilize climate, regulate the hydrological cycle, produce oxygen, store vast qualities of otherwise harmful carbon dioxide, and provide habitats for countless varieties of plants and animals, many of which are useful to human beings.
- *The cure-for-cancer argument*: Wilderness areas "contain the greatest source of medicinal natural resources."[15] If wildlands are destroyed, many potential new medicines may be forever lost.
- *The art gallery argument*: Wilderness areas often include places of great scenic beauty. Opening such areas to mining, logging, and other forms of development would mar or destroy such beauty.

---

14. Michael P. Nelson, "Rethinking Wilderness: The Need for a New Idea of Wilderness," *Philosophy in the Contemporary World* 3:2 (Summer 1996), p. 7.

15. Michael P. Nelson, "An Amalgamation of Wilderness Preservation Arguments," in Callicott and Nelson, eds., *The Great New Wilderness Debate*, p. 159.

- *The recreation argument*: Wilderness provides unique and incomparable opportunities for hiking, backpacking, canoeing, snowshoeing, horseback riding, fly-fishing, photography, and other outdoor recreational activities in remote, largely unspoiled locations.
- *The therapy argument*: Numerous studies have shown that time spent in nature can reduce stress, lower blood pressure, alleviate anxiety, improve physical fitness, and boost many other measures of physical and mental well-being.[16] As John Muir famously urged, "Climb the mountains and get their good tidings. Nature's peace will flow into you as sunshine flows into trees. The winds will blow their own freshness into you, and the storms their energy, while cares will drop off like autumn leaves."[17] In our increasingly urbanized, sedentary, and gadget-addicted world, the therapeutic benefits of wild nature may be more important than ever.
- *The classroom argument*: Time spent in wild places can teach humility, proper perspective, a sense of one's natural roots, respect for nature, responsible stewardship, and other important lessons about the natural world and our place in it.
- *The character-building argument*: Numerous writers have argued that "many of the attributes most distinctive of America and Americans are the impress of the wilderness."[18] Oft-admired qualities such as self-reliance, individualism, practicality, resourcefulness, and "hardihood, resolution, and the scorn of discomfort and danger"[19] are frequently cited as examples of such "pioneer" or "frontier" virtues. Long-distance hikers, hunters, backpackers, and others who encounter and struggle with wilderness often attest to its character-building effects.
- *The future generations argument*: Those who have experienced wild places of great scenic splendor and iconic wildlife are often profoundly grateful for those special and sometimes formative moments. Many believe that

---

16. See Richard Louv, "Outdoors for All," *Sierra Magazine*, 104:3 (May/June 2019), pp. 27–28.

17. John Muir, *Our National Parks* (Boston, MA: Houghton Mifflin, 1901), p. 56.

18. Aldo Leopold, "Wilderness as a Form of Land Use," *Journal of Land and Public Utility Economics* 1:4 (October 1925); reprinted in Curt Meine, ed., *Aldo Leopold:* A Sand County Almanac *and Other Writings on Ecology and Conservation* (New York: The Library of America), p. 281. The most famous proponent of this so-called frontier thesis was the distinguished historian Frederick Jackson Turner (1861–1932). For a discussion of Turner's significance in American environmental thought, see Nash, *Wilderness and the American Mind*, pp. 145–47.

19. Theodore Roosevelt, *Outdoor Pastimes of an American Hunter* (New York: Charles Scribner's Sons, 1906), p. 375.

foreclosing such "bucket-list" experiences to future generations would be a grave act of intergenerational injustice.[20]

Each of these human-centered arguments has some persuasiveness, and collectively they make a powerful case for wilderness preservation. But the deepest reasons for preserving wild places are biocentric rather than anthropocentric. They require a shift from what Pope Francis calls a language of "masters, consumers, [and] ruthless exploiters" to one of "fraternity and beauty."[21]

Let's consider three biocentric arguments for wilderness preservation. The first centers on animal welfare, the second on biodiversity, and the third on respect for nature.

In previous chapters we saw that all living things have intrinsic value and moral standing, and that some life-forms—especially sentient animals—have greater moral status than others. When wilderness areas are logged, mined, flooded, converted to farmland, or otherwise damaged or destroyed for human benefit, many sentient animals typically suffer and die. For example, when large sections of the Amazon rainforest are burned to clear space for cattle ranching, innumerable sentient creatures suffer, lose their homes, or are killed. Sometimes, as we saw, such losses may be justified by overriding human benefits, but the moral concern and respect that we should have for animal welfare creates a presumption in favor of wilderness preservation.

Second, wilderness plays a crucial role in preserving biodiversity. According to a 2019 U.N. report, up to a million of the approximately eight million plant and animal species on earth are at risk of extinction, many within the next few decades.[22] Well over half of all land-based biodiversity is found in the tropics, much of it in wilderness or lightly populated areas. In North America, wilderness areas provide critical habitat for wolves, bears, cougars, elk, mountain sheep, and other large wildlife species. As we shall see in the following chapter, there are strong anthropocentric reasons for being concerned with biodiversity loss.[23] Many important food crops, for example, depend on insect pollinators

20. Other notable anthropocentric arguments for wilderness preservation include the laboratory argument (wilderness provides unique opportunities for scientific research), the cathedral argument (wilderness can be a locus of spiritual, religious, or transcendent experiences), and the storage silo argument (wilderness is a vast and largely untapped storage vault of potentially useful genetic information).

21. Pope Francis, *Laudato Si'* (May 24, 2015), § 11 (online). Web. 18 March 2020.

22. Bill Chappell and Nathan Rott, "One Million Plant and Animal Species Are at Risk of Extinction, U.N. Report Says," *NPR Research News*, May 6, 2019. Web. 20 March 2020.

23. This is a major reason why some critics of the received view of wilderness support protected wilderness areas so long as they are reconceived as "biodiversity reserves."

whose populations are now in sharp decline. The importance of preserving a rich abundance of diverse life-forms is even more obvious from a biocentric perspective. If life itself has value, then a bountiful variety of life should be valued all the more.

The final biocentric argument for wilderness preservation turns on the values that would be lost in a thoroughly human-dominated world. Civilization and its human values are undoubtedly good things, but are they the only good things? Does an eagle's flight, a wolf's howl, a trout's leap, and a meadowlark's liquid song count for nothing in the balance? Humans have already destroyed most of wild nature. Have we a right to destroy what remains? To say that we do would betray a profound lack of respect for nonhuman nature, emblematic of what Paul Taylor calls a "directly exploitative attitude"[24] toward the natural world. For billions of years, wild nature has been the womb of biotic value and generativity on planet Earth. It is our ancestral home, the crucible on which human nature and civilization have been hammered out as steel on an anvil. We must agree, therefore, with Aldo Leopold that "while the reduction of wilderness has been a good thing, its extermination would be a very bad one."[25] In our relentless quest to convert earth's bounties to profits, pleasure, and convenience we can afford to set aside some portions of the globe where untrammeled nature may continue to exist wild and free. Once lost, wilderness can never be restored to its original state. The cost of failing to protect what little remains of wilderness—the *value* cost—is too high to pay.

In sum, there seem to be both powerful anthropocentric and nonanthropocentric reasons for wilderness preservation. But what can be said on the other side of the ledger? Can an even stronger case be made *against* wilderness preservation? Let's consider the "con" side of "the great wilderness debate."

11.3 The Case against Wilderness

The current wilderness debate began in the late 1980s with the publication of Indian author Ramachandra Guha's influential article, "Radical American

---

See, for example, J. Baird Callicott, "Should Wilderness Areas Become Biodiversity Reserves?" in Callicott and Nelson, eds., *The Great New Wilderness Debate*, pp. 585–94.

24. Paul W. Taylor, *Respect for Nature: A Theory of Environmental Ethics* (Princeton, NJ: Princeton University Press, 1986), pp. 273–74.

25. Quoted in Curt Meine, *Aldo Leopold: His Life and Work* (Madison, WI: University of Wisconsin Press, 1988), p. 245.

Environmentalism and Wilderness Preservation: A Third World Critique."[26] Two years later, J. Baird Callicott extended Guha's arguments into a full-blown critique of what Callicott termed the "received view" of wilderness.[27] This was followed, in 1995, by William Cronon's much discussed attack on the "myth" of wilderness in his essay, "The Trouble with Wilderness, or Getting Back to the Wrong Nature."[28] Callicott and Michael P. Nelson co-edited two hefty anthologies on the ensuing controversy in 1998 and 2008, respectively.[29] Most recently, Mark Woods has authored a judicious critical overview of the debate in his 2017 book, *Rethinking Wilderness*.[30]

Recent critics of wilderness come at it from different angles. Some argue against *any* concept of wilderness, others only against the received concept. Similarly, some argue against any practice of wilderness preservation, some only against current practices. We have already considered and rejected two common anti-wilderness arguments: (1) the claim that the received view of wilderness creates an illegitimate separation of humans from nature;[31] and (2) the contention that the concept of wilderness preservation is incoherent, because any attempt to preserve wilderness in an unchanging condition would "trammel" it and thus transform it into non-wilderness.[32] Now let's consider five additional arguments against wilderness preservation.

---

26. Ramachandra Guha, "Radical American Environmentalism and Wilderness Preservation: A Third World Critique," *Environmental Ethics* 11:1 (1989), pp. 71–83.

27. J. Baird Callicott, "The Wilderness Idea Revisited: The Sustainable Development Alternative," *The Environmental Professional* 13:3 (1991), pp. 235–47; reprinted in Callicott and Nelson, eds., *The Great New Wilderness Debate*, pp. 337–66.

28. William Cronon, "The Trouble with Wilderness, or Getting Back to the Wrong Nature," in William Cronon, ed., *Uncommon Ground: Toward Reinventing Natur*e (New York: W. W. Norton, 1995), pp. 69–90; reprinted in Callicott and Nelson, eds., *The Great New Wilderness Debate*, pp. 471–99. Citations to this article will be to the Callicott and Nelson reader.

29. Callicott and Nelson, eds., *The Great New Wilderness Debate,* and *The Wilderness Debate Rages On: Continuing the Great New Wilderness Debate* (Athens, GA: University of Georgia Press, 2008).

30. Woods, *Rethinking Wilderness*.

31. See Callicott, "The Wilderness Idea Revisited," pp. 348–51; Cronon, "The Trouble with Wilderness," pp. 484–85; and Nelson, "Rethinking Wilderness," p. 7.

32. Callicott, "The Wilderness Idea Revisited," pp. 353–54; and Nelson, "Rethinking Wilderness," p. 7.

The No-Wilderness Argument

Some critics argue that the whole concept of wilderness is a myth, because the so-called "wilderness" that the first European settlers encountered was actually a landscape significantly modified by native peoples. Moreover, no true wilderness exists today, because every part of the globe has been affected by atmospheric pollution, nuclear fallout, human-caused climate change, and other human impacts.[33]

Like the argument that wilderness preservation is impossible because wilderness isn't static and cannot possibly be preserved unchanged, this argument assumes a "purist" conception of wilderness that no informed environmentalist actually accepts. While it is certainly a gross exaggeration to say that *no* unmodified, pristine wild places existed in North America at the time of the European discovery,[34] it is widely known that native peoples modified the land by ax, arrow, and fire. The Wilderness Act does not assume any purist definition of "wilderness." On the contrary, it speaks of landscapes that "have been affected *primarily* by the forces of nature, with the imprint of man *substantially unnoticeable*."[35] Moreover, subsequent congressional designations of wilderness—many in areas fairly heavily impacted by human activity—make clear that Congress does not accept any purist definition either. The claim that wilderness exists only if it is completely unmodified and pristine does not comport with either ordinary or legal usage; it is a tendentious notion concocted to make wilderness an easy target for attack.

The Ill-Gotten Gains Argument

Many critics of wilderness charge that the very concept of wilderness is morally tainted.[36] In several notable cases, wilderness preservation in the United

---

33. Callicott, "The Wilderness Idea Revisited," pp. 351–53; Cronon, "The Trouble with Wilderness," pp. 487; and Nelson, "Rethinking Wilderness," pp. 6–7. See generally Bill McKibben, *The End of Nature* (New York: Random House, 1989).

34. See Holmes Rolston, "The Wilderness Idea Reaffirmed," *The Environmental Professional* 13 (1991), pp. 376–79 (available online). Rolston rightly notes, for example, that there is little evidence of native people's impacts in many high-mountain terrains and arid deserts.

35. The Wilderness Act of 1964, reprinted in Callicott and Nelson, eds., *The Great Wilderness Debate*, p. 121. Emphasis added.

36. See, for example, Charles E. Kay, "Afterword: False Gods, Ecological Myths, and Biological Reality," in Charles E. Kay and Randy T. Simmons, eds., *Wilderness and Political Ecology: Aboriginal Influences and the Original State of Nature* (Salt Lake City, UT: University of Utah Press, 2002), p. 259.

States was made possible only by killing or dispossessing indigenous peoples.[37] According to these critics, wilderness is thus what lawyers call "a fruit of the poisonous tree," rendering the preservation of wilderness areas today morally compromised.

As Mark Woods points out, this argument is faulty in both its factual claims and its logic. "With few exceptions," Woods notes, "the intent behind the killing and removal of Native American Indians was not to make way for wilderness areas. Rather, in almost all instances they were killed or removed to make way for railroads, mines, livestock grazing, homesteads, farms, timber operations, water developments, cities and the like."[38] Moreover, if wilderness areas should be abandoned because of their morally tainted origins, then so too should shopping malls, sports stadiums, hospitals, and universities, which in many cases were also built on indigenous lands.[39] Since nobody seriously proposes the latter, wilderness should not be treated differently.[40]

The Social Constructivist Argument

Some critics of wilderness have argued that the concept of wilderness is a cultural invention, a socially constructed idea.[41] This provides the key premise of what Mark Woods labels the social constructivist argument against wilderness. This is how Woods formulates the argument:

1. In order for the concept of wilderness to make sense, it must connote the idea of nature as existing independent of human cultures.
2. The concept of wilderness thus presupposes that a meaningful conceptual distinction can be made between human cultures and nonhuman nature.
3. Because wilderness and nature, like all other concepts, are human social constructions, it is problematic to say that wilderness exists independent from human cultures. That is, because the ideas of nature and wilderness are socially constructed, there are no

37. For examples, see Woods, *Rethinking Wilderness*, pp. 123–25. What I call the ill-gotten gains argument, Woods variously dubs "the imperial argument" (*Rethinking Wilderness*, p. 125) or "the moral argument." Woods, "Wilderness," in Dale Jamieson, ed., *A Companion to Environmental Philosophy* (Malden, MA: Blackwell, 2001), p. 356.

38. Woods, "Wilderness," p. 357.

39. Woods, *Rethinking Wilderness*, pp. 141–42.

40. As Woods notes, the ill-gotten gains argument might support some less extreme conclusion, such as the need for some form of restitutive or restorative justice to native peoples. Woods, *Rethinking Wilderness*, pp. 143–44.

41. See, for example, Cronon, "The Trouble with Wilderness," p. 483.

non-socially constructed natural areas that exist independent of human cultures.

4. What the concept of wilderness connotes—the idea of nature as existing independent of human cultures—is non-existent.

Conclusion: Thus, the concept of wilderness is flawed.[42]

This postmodernist argument trades on a confusion. As Holmes Rolston notes, the fact that many aboriginal and some modern cultures do not have a word for wilderness does not show that wilderness does not exist independent of human knowers and language-users.[43] If that were the case, then the same could be said about recent scientific concepts like DNA and neutron star, which clearly denote realities that exist outside the human mind. To say that there is no wilderness because the *concept* of wilderness is a social construct is to confuse the *lens* through which we see reality with reality itself. Put differently, the fact that human concepts and categorizations mediate and condition what we can experience, does not imply that there are no objective or mind-independent objects and states of affairs.[44] Humans did not make wilderness; wilderness made humans.

The Environmental Justice Argument

Probably the strongest criticism of wilderness preservation is that it is elitist and fails to address many of the most urgent environmental problems that exist today. Guha made this charge in the much-discussed 1989 paper[45] that kicked off the current wilderness debate, and it has been repeated and amplified by many critics since. The charge is really twofold. First, wilderness preservation is alleged to be elitist because wilderness is mostly enjoyed by "yuppie backpackers" and other affluent and privileged people; and second, that wilderness advocates tend to exaggerate the importance of wilderness preservation while largely ignoring more pressing environmental problems such as climate change, pollution, nuclear proliferation, overconsumption, and the unjust imposition of environmental burdens on people of color and the poor. As Guha noted, wilderness preservation isn't a real option in many areas in the world, because little

42. Woods, *Rethinking Wilderness*, p. 66.

43. Holmes Rolston III, *A New Environmental Ethics: The Next Millennium for Life on Earth* (New York: Routledge, 2012), p. 177.

44. Woods, *Rethinking Wilderness*, pp. 79–80. For a brief and witty critique of the postmodernist notion that reality is socially constructed, see Alvin Plantinga, *Warranted Christian Belief* (New York: Oxford University Press, 2000), pp. 422–37.

45. Guha, "Radical American Environmentalism," pp. 71–83.

MAKING A DIFFERENCE

Dr. Robert D. Bullard: The Father of Environmental Justice

Beginning in the U.S. South in the early 1980s, the modern **environmental justice** movement has fought to end **environmental racism** and ensure that minority populations and low-income populations have an equal voice in decision-making processes that affect the environmental quality of their neighborhoods and communities. For more than a quarter-century, the leader of the environmental justice movement has been Dr. Robert ("Bob") Bullard.

Bullard was born in Elba, Alabama in 1946 and educated at two historically black southern colleges, before receiving his PhD in sociology in 1976. Over the course of his long and distinguished teaching career, Bullard has taught at Texas Southern University; the University of Tennessee; the University of California, Riverside; and Clark Atlanta University. He is currently Distinguished Professor of Urban Planning and Environmental Policy at Texas Southern University in Houston, Texas.

Author of eighteen books, Bullard first came to prominence in the early 1990s when he published two influential books on environmental justice: *Dumping in Dixie: Race, Class, and Environmental Quality* (1990) and *Unequal Protection: Environmental Justice and Communities of Color* (1994).

In those and other works, Bullard demonstrated convincingly that minorities and low-income populations bear a disproportionate share of environmental risks and burdens. They are far more likely than whites to breathe polluted air; live near hazardous or noxious landfills, refineries, sewage plants, or incinerators; and more likely to suffer health effects from agricultural herbicides and pesticides. They are also less likely to have the resources or the political clout to protect themselves from such environmental hazards.

Bullard and other activists in the environmental justice movement have been successful in raising awareness of issues of environmental racism and environmental justice. In 1992 the Environmental Protection Agency (EPA) created the Office of Environmental Justice. Two years later, President

(*Continued*)

Clinton signed an executive order directing federal agencies to address issues of environmental justice for minority and low-income populations. Beginning in 2014, problems with lead-contaminated drinking water in predominantly African-American Flint, Michigan created outrage and charges of environmental racism. And in the 2020 American presidential election campaign, all the leading Democratic candidates proposed ambitions plans for addressing both climate change and environmental justice.

For his tireless efforts on behalf of environmental justice, Dr. Bullard has received numerous awards and accolades, including the Sierra Club's John Muir Award in 2013 and the William Julius Wilson Award for the Advancement of Justice.

or no wilderness still remains there. As he saw it, a strong focus on wilderness preservation reflects a distinctly American point of view that has little relevance to many other parts of the world.

Guha's charge was directed mainly at 1980s-era deep ecologists, who did often stress wilderness preservation while giving short shrift to other serious environmental problems, especially in the developing world. But though there is clearly some truth to Guha's charge, the conclusion he draws—that efforts to preserve wilderness should be abandoned or curtailed—does not follow. Classical music concerts, sailboat races, ski resorts, and professional tennis tournaments cater mostly to the well-to-do. Does it follow that they should be banned or restricted? If the poor and underprivileged have too few opportunities to enjoy wild nature, then we should consider ways to make those opportunities more widely available, not curtail them for all.[46]

The charge that wilderness advocates attach undue importance to wilderness preservation—to the exclusion of other important ecological concerns—is a fair criticism of *some* wilderness defenders. But wilderness advocacy should not be confused with single-minded or obsessive wilderness advocacy. As Woods notes, most mainstream environmental groups since the 1960s have supported a wide range of environmental causes, not simply wilderness preservation.[47]

---

46. Louv, "Outdoors for All." The movement supports universal and equitable access to nature as a basic human right.

47. Woods, *Rethinking Wilderness*, pp. 205–6.

The Economic Argument

Most opposition to wilderness preservation comes, in fact, not from environmentalists and left-leaning academics but from corporate interests and developers. When wild places are designated as wilderness they are closed to most commercial activities and residential or infrastructure development, though various exceptions are permitted. There is thus frequently an economic cost to wilderness preservation. Some critics claim that when wilderness and economic interests clash, economic interests should normally prevail.

This argument, even if sound, won't exclude all wilderness preservation efforts, because some wilderness areas have little economic value. But a deeper problem with the argument is that it views nature from an anthropocentric and excessively economic point of view. As we've seen, there are strong biocentric reasons for protecting wilderness. Moreover, the Wilderness Act explicitly recognizes that wild places may be set aside for the sake of their "ecological, geological, or other features of scientific, educational, scenic, or historical value." Allowing economic considerations to trump all other forms of value is inconsistent with the biocentric reasons that support wilderness preservation. Thus, while it certainly makes sense to weigh the economic costs of wilderness protection, especially when such costs are high, the human and biocentric values underlying wilderness preservation preclude viewing economic considerations as paramount.

In the final balance, the arguments for wilderness preservation seem to be much stronger than those opposing it. Setting aside scenic wild places for all future generations to enjoy was indeed one of the best ideas America ever had. May such sanctuaries long remain wild and free.

Chapter Summary

1. This chapter examines the notion of wilderness and how important it is to preserve remaining areas of wilderness. The term "wilderness" is used in various senses, and its meanings have shifted over time. *The Merriam-Webster Dictionary* defines "wilderness" as "a tract or region that is uncultivated and uninhabited by human beings." Under the landmark Wilderness Act of 1964, "wilderness" is defined for purposes of federal law as a sizeable area of public land "where the earth and its community of life are untrammeled by man, where man himself is a visitor who

does not remain." Some critics argue that the whole idea of wilderness is flawed and should be jettisoned, but so long as we avoid unduly "purist" conceptions of wilderness, the idea seems viable and important.

2. There are both anthropocentric and biocentric arguments for wilderness preservation. Anthropocentric reasons include ecological, recreational, pharmacological, aesthetic, therapeutic, character-building, and perspective-giving benefits of unspoiled wild places. Biocentric reasons include the value of wilderness in promoting animal welfare and preserving biodiversity, as well as in promoting and preserving various forms of intrinsic natural value that merit human care and respect. Cumulatively, those arguments seem to make a strong case for wilderness preservation.

3. Critics of wilderness attack it from many angles. Some claim that the concept of wilderness is an empty placeholder because there aren't any genuinely "pristine" areas of nature left in the world. Others argue that the idea of wilderness rests on an illegitimate bifurcation of humans and nature, since Darwinism has taught us that humans are fully part of nature. Others claim that the concept of wilderness preservation is self-contradictory because any attempt to "preserve" wilderness from change would "trammel" it with human management and control. Still other critics attack the concept of wilderness as morally tainted or as a mere social construction. Influential critic Ramachandra Guha argues that the ideal of wilderness preservation is elitist, unduly focused on the American environmental context, and based on misplaced environmental priorities that have little relevance to the urgent ecological problems of the Third World. Finally, some critics argue that wilderness preservation can cost jobs and stunt economic growth, and should usually be nixed for those reasons. It was seen that none of these criticisms of wilderness or wilderness preservation is convincing.

Discussion Questions

1. What anthropocentric reasons are there for wilderness preservation? Cumulatively, how strong are these reasons?

2. What biocentric reasons are there for wilderness preservation? Taken together, how strong are those considerations?

3. What are Callicott's objections to the "received view" of wilderness? Are they persuasive?
4. What are the main arguments against wilderness preservation? Are they convincing?
5. Is concern for wilderness preservation elitist? Is it biased in favor of the young, fit, and affluent? Does it reflect a distinctly American form of environmentalism that may not apply to many parts of the globe?
6. In general, which is more important: economic growth or wilderness preservation? Should wilderness only be preserved when doing so does not have high economic costs?

Further Reading

For a classic work on American attitudes to wilderness, see Roderick Frazier Nash, *Wilderness and the American Mind*, 5th ed. (New Haven, CT: Yale University Press, 2014). For an equally classic history of ecology, see Donald Worster, *Nature's Economy: A History of Ecological Ideas*, 2nd ed. (Cambridge: Cambridge University Press, 1994). For a wide range of readings, pro and con, on wilderness preservation, see J. Baird Callicott and Michael P. Nelson, eds., *The Great New Wilderness Debate* (Athens, GA: University of Georgia Press, 1998); and J. Baird Callicott and Michael P. Nelson, eds., *The Wilderness Debate Rages On: Continuing the Great New Wilderness Debate* (Athens, GA: University of Georgia Press, 2008). For a compact overview of arguments for and against wilderness preservation, see Mark Woods, "Wilderness," in Dale Jamieson, ed., *A Companion to Environmental Philosophy* (Malden, MA: Blackwell, 2001), pp. 349–61. For a fuller and more recent treatment, see Mark Woods, *Rethinking Wilderness* (Peterborough, Ontario: Broadview Press, 2017). For an influential defense of the idea of wilderness, see Holmes Rolston III, "The Wilderness Idea Reaffirmed," *The Environmental Professional* 13 (1991), pp. 370–77, available online (along with many other works by Rolston) at the Rolston Digital Archives.

Chapter 12

The Extinction Crisis

Extinctions are nothing new in nature. Over 98 percent of all species that ever existed on earth are now extinct. Over the past 500 million years, at least five great mass extinctions have occurred. The largest was the end-Permian extinction, which occurred roughly 252 million years ago. Nearly 90 percent of all plant and animal species perished then, probably as a result of rapid global warming.[1] A more famous mass extinction occurred 65 million years ago when the dinosaurs died out, perhaps due to a large asteroid strike that plunged the earth into a prolonged period of darkness and cold. After each of these mass extinctions, it took at least 20 million years for previous levels of biodiversity to be restored.[2] Today, in the human-dominated geological epoch we call the Anthropocene, a sixth great extinction is occurring—one caused almost entirely by us. Normally in nature, roughly one to five species become extinct each year. Now, dozens, if not hundreds, of species of plants and animals are vanishing every single day. And because of factors such as habitat destruction, climate change, overharvesting, pollution, and invasive species, the extinction crisis is rapidly worsening. According to a recent landmark U.N. report, about a million species are at risk of extinction over the next few decades.[3] Worldwide, humans have wiped out 60 percent of all mammals, birds, insects, and reptiles since 1970.[4] A quarter of all mammals and a third of all amphibians are now listed as "threatened." In this chapter, we explore why biodiversity matters and how, in general terms, we should respond to the extinction crisis.

---

1. Elizabeth Kolbert, *The Sixth Extinction: An Unnatural History* (New York: Henry Holt and Co., 2014), p. 103.

2. Edward O. Wilson, *The Diversity of Life* (New York: W. W. Norton, 1993), p. 31.

3. U.N. Intergovernmental Science-Policy Platform on Biodiversity and Ecological Services (IPBES), "Global Assessment Report on Biodiversity and Ecological Services," Ipbes.net, May 6, 2019. Web. 22 March 2020.

4. Damian Carrington, "Humanity Has Wiped Out 60% of Animal Populations since 1970," *The Guardian*, October 29, 2018. Web. 24 March 2020.

12.1 Why Species Preservation Is Important

Nobody knows how many species of plants and animals now exist on earth, but common estimates range from five to thirty million.[5] By this measure of biodiversity—what scientists call species richness—the earth is blessed with an amazing variety of life-forms.[6] How wondrous that we live in a world that includes such beautiful and fascinating creatures as Bengal tigers, snow leopards, giraffes, African elephants, bald eagles, ruby-throated hummingbirds, monarch butterflies, dolphins, and blue whales! How concerned should we be about biodiversity loss? Extremely concerned, for lots of reasons.

Arguments for preserving natural variety are of two sorts: anthropocentric and nonanthropocentric. Anthropocentric arguments point to benefits that humans derive from a particular species or collection of species. Nonanthropocentric reasons focus either on the intrinsic value of species, or on the way a species contributes to ecosystem health. Let's begin with some anthropocentric reasons for species preservation.

Many Species Are Directly Useful to Human Beings

Nature, of course, provides common human food sources such as corn, wheat, rice, cows, fish, chickens, fruits, and nuts. Trees provide timber, purify the water, prevent erosion, produce oxygen, and slow global warming by storing carbon that would otherwise wind up in the oceans or the atmosphere. Many modern medicines and drugs are derived from plants or animals. Bees and other wild pollinators play a vital role in world food production. Worms help fertilize the soil. Bats and birds help control gnats and mosquitoes. Nor should we forget what naturalist E. O. Wilson calls "the little things that run the world." Without bacteria, fungi, and other tiny decomposers of organic matter, Wilson claims, the planet would literally rot in stinking piles of biotic debris, and nearly all life on earth would become extinct within a few decades.[7]

---

5. Holmes Rolston III, *A New Environmental Ethics: The Next Millennium for Life on Earth* (New York: Routledge, 2012), p. 144. Only about 1.6 million species have been identified to date. Since only about 13,000 to 15,000 new species are identified each year, we are nowhere close to discovering all species.

6. There are many different ways to measure biodiversity, including the number of different species in a given area (species richness) and the range of genetic diversity amongst species in that area (taxonomic diversity). One can also speak of diversity *within* a given habitat (alpha diversity) or *between* two habitats (beta diversity).

7. Edward O. Wilson, "The Little Things that Run the World," *Conservation Biology* 1:4 (December 1987), p. 345.

Many Species Are Indirectly Useful to Humans

Some species aren't of immediate use to humans, but they do benefit us indirectly. Consider plankton, the tiny plant, animal, and bacterial organisms that drift through the world's oceans and lakes in vast numbers. Humans don't eat plankton, but many aquatic creatures do. Plankton, in fact, is the foundation of the ocean food chain. As such, it provides an incalculable, indirect benefit to humans. Acorns provide another example of indirect benefit to humans. Though rarely consumed by humans, acorns provide essential food for deer, wild turkeys, ducks, and other animals that are useful to our species. Because it is an ecological law that diversity begets diversity, far more species may be of indirect benefit to humans than one might imagine.[8]

Some Species, If Preserved, Might Someday Be Useful to Human Beings

This is the so-called cure-for-cancer argument for species preservation. As noted, a great many medicines and drugs are made from plants and animals. Living things are chemical factories and libraries of genetic information that have accumulated over countless eons. As Holmes Rolston notes, "destroying species is like tearing pages out of an unread book, written in a language humans hardly know how to read, about the place where we live."[9] Only a small percentage of species have been tested for their potential benefits to humans. Who knows what miraculous cures or life-saving information might be gleaned by preserving nature's storehouse of genetic and chemical information?

Species Enrich Human Life in Numerous Other Ways as Well

Apart from their practical use or utilitarian "cash value," wild and domesticated species of plants and animals can benefit humans in numerous other ways. Pets can improve health and be sources of love, comfort, and companionship. Wild, biotically rich places provide recreational opportunities, such as hiking, backpacking, hunting, fishing, and bird-watching. Flowering trees, colorful songbirds and butterflies, fleet antelope, soaring ospreys, majestic redwoods, high-flying cranes, alpine wildflowers, and many other organisms are beautiful and thus sources of aesthetic appreciation, wonder, and emotional uplift. Some species have religious, symbolic, historical, or cultural value, as grizzly bears and bison did (and do) for the Plains Indians of North America or as the bald eagle

---

8. Bryan G. Norton, *Why Preserve Natural Variety?* (Princeton, NJ: Princeton University Press, 1987), p. 61.

9. Rolston, *A New Environmental Ethics*, p. 131.

does in the United States today. There is growing evidence that "forest bathing" and other forms of "nature therapy" can lower blood pressure, reduce stress, and have other health benefits.[10] Many nature enthusiasts, such Aldo Leopold[11] and William James,[12] claim that hunting, canoeing, and other sorts of strenuous outdoor activities can build character and moral fiber. And for scientists and naturalists, nature and wild species provide endless opportunities for scientific investigation, discovery, and understanding.

The foregoing are all human-centered reasons for species preservation. But as we've seen in earlier chapters, we cannot focus solely on human concerns in our thinking about the environment. Nonhuman organisms, species, and ecosystems have value for their own sakes, and sometimes those values trump competing human interests. Consequently, we must consider both anthropocentric and nonanthropocentric grounds for concern with biodiversity loss. Frequently, such concerns are rooted in religious or spiritual beliefs, such as **animism** (the view that many or all natural objects, as well as living things, have a mind or soul) or the common Judeo-Christian doctrine that all creatures "give glory to God by their very existence."[13] Since this is a philosophy textbook, not a work of theology, we will consider only nonreligious, nonanthropocentric arguments for species preservation. I will suggest three.

When a Species Becomes Extinct, Something of Irreplaceable Value Is Forever Lost

As noted earlier, environmentalists are usually more concerned with the good of a species than they are with the good of individual members of that species. In 1983, for example, a bison fell through the ice while crossing the Yellowstone River, and was allowed to drown by park officials. Those officials defended the decision by claiming that this was a case where it was best to "let nature take its course."[14] The following year, however, Yellowstone park rangers rescued a sow

10. Qing Li, "Forest Bathing Is Great for Your Health: Here's How to Do It," *Time Magazine*, May 1, 2018. Web. 15 May 2020.

11. Aldo Leopold, *A Sand County Almanac with Essays on Conservation from Round River* (New York: Ballantine Books, 1970), pp. 269–70.

12. William James, "The Moral Equivalent of War," in Bruce Kuklick, ed., *William James, Writings 1902–1910* (New York: The Library of America, 1987), pp. 1281–93.

13. Pope Francis, *Laudato Si'* (May 24, 2015), §33 (available online). Web. 28 March 2020. Cf. St. Augustine, *The City of God*, translated by Marcus Dodds (New York: The Modern Library, 1950), Book 12, chap. 4 ("it is not with respect to our own convenience or discomfort, but with respect to their own nature, that creatures are glorifying to their Artificer").

14. Rolston, *A New Environmental Ethics*, pp. 71–72.

grizzly and her three cubs when they were trapped on an island in Yellowstone Lake and faced imminent starvation. In this case, the rangers claimed, concern for an endangered species overrode the usual policy of noninterference with natural processes. Most environmentalists would agree with these contrasting decisions because in their eyes species have greater value than individual organisms. Why? Partly because the death of a species is nearly always more consequential than the death of an individual plant or animal. As Holmes Rolston notes, when an individual plant or animal dies, a single fleeting life is cut short, while others of its kind continue to exist. Extinction, by contrast, completely shuts down a generative process and ends a wondrous story of triumph and tragedy that has gone on for eons.[15] The species is a general pattern, a dynamic form, which has for ages successfully navigated the crucible of evolution and defended a particular form of life. As long as the pattern survives, the story continues. When the pattern is snuffed out, the evolutionary saga ends forever.

Biodiversity Loss Can Cause "Extinction Cascades" That Can Spell Ecological Disaster

One of the central tenets of modern ecology is summed up in Barry Commoner's first law of ecology: "Everything is connected to everything else."[16] All living things exist in complex webs of mutual dependence. As a result, if one species goes extinct, other species that interact with it with may also be threatened.[17] One prominent biologist, Peter Raven, estimates that for each plant that becomes extinct, a dozen other species of plants and animals can be expected to follow.[18] The upshot is that extinctions can snowball and quickly lead to serious biodiversity loss, or even to ecosystem collapse. A well-known example is the Escudilla Wilderness Area of eastern Arizona, profiled in Aldo Leopold's famous essay, "Thinking Like a Mountain." Leopold noted that when wolves were extirpated from Escudilla Mountain, it caused a spike in the deer population, which led in turn to extreme overgrazing and a severely degraded ecosystem. Another example is when sea otters were eliminated from most of the California Pacific coast in the early nineteenth century, leading to an explosion of sea urchins and major damage to the coastal kelp bed ecosystems,

15. Rolston, *A New Environmental Ethics*, p. 135.

16. Barry Commoner, *The Closing Circle: Nature, Man, and Technology* (New York: Alfred A. Knopf, 1971), p. 33.

17. See University of Exeter, "Biodiversity Loss Raises Risk of 'Extinction Cascades,'" *ScienceDaily*, February 19, 2018. Web. 30 March 2020.

18. Holmes Rolston III, *Environmental Ethics: Duties to and Values in the Natural World* (Philadelphia, PA: Temple University Press, 1988), p. 130.

with profound cascade effects on other forms of marine life.[19] To be sure, nature usually has enough resilience and redundancy to bounce back from the loss of one or a few species. But both nature and humanity can pay a high price when we forget that in nature "everything is connected."

Biodiversity Itself Is of Inherent Value

Species, ecosystems, and individual organisms all have intrinsic value—that is, value for their own sakes—and so too does the state or condition of biodiversity itself. We rightly prize variety in art, music, literature, cuisine, entertainment, and scenery. The same is true of biology. A great abundance, variety, and complexity of life-forms permits many variegated forms of excellence and beauty to exist—from the keen eyesight of the golden eagle, to the brilliant plumage of the scarlet tanager, to the echolocation of bats, to the speed of a cheetah.[20] Biodiversity also allows more life to exist in more places, since diverse organisms are able to find niches and colonize different parts of the land and oceans. Finally, biodiversity is inherently valuable because it boosts nature's resiliency, allowing plant and animal populations to bounce back more quickly when mass extinctions or other major ecological disturbances occur, such as the loss of billions of American chestnut trees from the eastern United States in the early decades of the twentieth century due to a fungus blight. Because America's hardwood forests were so diverse, oaks, maples, mountain laurels, and other plants were able to quickly fill in the places in the forest canopy where the stately chestnut trees had long reigned as monarchs of the forest.[21]

Together, the foregoing arguments make a strong case for urgent concern and bold action in combating the rapidly unfolding calamity of mass extinction in the Anthropocene. But do they require that we try to save *all* species, regardless of expense, effort, likelihood of success, or costs to human interests? Do they imply, as E. O. Wilson claims, that "we should judge every scrap of biodiversity as priceless" and never "knowingly allow any species or race to go

---

19. Malcolm L. Hunter Jr. and James P. Gibbs, *Fundamentals of Conservation Biology*, 3rd ed. (Malden, MA: Blackwell, 2007), p. 78.

20. The idea that creation is better and more perfect if it contains diverse levels and kinds of beings is an old idea stretching back to the Greek Neoplatonists, and a commonplace in medieval thought. For a classic study of this long-influential idea, see Arthur O. Lovejoy, *The Great Chain of Being: A Study of the History of an Idea* (Cambridge, MA: Harvard University Press, 1936).

21. Which isn't to say that the chestnut blight wasn't an environmental calamity. The nuts were a major food source for people as well as for animals, and the lumber from chestnuts was highly prized.

MAKING A DIFFERENCE

Dr. Jane Goodall: Conservationist Icon

Scientist, conservationist, and tireless advocate for animal welfare, Dr. Jane Goodall is known throughout the world for her work with chimpanzees and her efforts to raise environmental awareness.

Goodall was born in London in 1934. Interested in animals from a young age, she traveled to Kenya in 1957. There she met Louis Leakey, the famed paleontologist. Leakey hired her as his secretary and in 1960 arranged for her to study wild chimpanzees in Gombe Stream National Park in Tanzania.

Goodall spent years there closely observing chimpanzee behavior. Though the chimps at first ran away from Goodall, they eventually accepted her into their troop. Goodall was the first naturalist to prove that humans were not the only animals that use tools. She also surprised many in the scientific community by showing that chimps had distinct personalities, complex emotions, and close-knit social bonds, and that they sometimes hunted and ate meat.

Though Goodall had never attended college previously, she was admitted to graduate school at Cambridge University in 1962 and received a PhD in ethology (the science of animal behavior) in 1965.

Author of more than twenty books and featured in numerous documentaries and magazine articles, Goodall is now one of the most admired conservationists and spokespersons for animal welfare in the world. In 1977 she founded the Jane Goodall Institute, dedicated to protecting chimpanzees and promoting sustainable conservation practices in Africa. In 1991, she co-founded Roots and Shoots, a global leadership program for youth, which now has chapters in over a hundred countries.

For her work on animal behavior, conservation, and animal welfare, Goodall has won countless awards and been awarded numerous honorary degrees. In 2004 she became Dame Jane Morris Goodall DBE when Queen Elizabeth named her Dame Commander of the Most Excellent Order of the British Empire. In 2019 *Time* magazine named her one of the one hundred most influential people in the world. Now in her eighties, Goodall refuses to slow down. She regularly travels hundreds of days a year spreading the gospel of conservation and care for the natural world.

extinct."[22] Such rhetoric is inspiring, but is it realistic? The sad truth is that biodiversity loss is now so severe that we lack the human and financial resources to preserve all species, and many species are doomed no matter what heroic efforts we make to save them. By tragic necessity, therefore, we have no choice but to "play God" and establish what the U.S. Fish and Wildlife Service calls "priority rankings" for endangered and threatened species. But how should such rankings be made?

This critical issue is far too complex to be discussed in detail here. But some general suggestions may be helpful.

Some criteria for prioritizing species for preservation and recovery efforts are fairly obvious and uncontroversial. Thus, most environmentalists would agree that we should consider:

- The *magnitude* and *immediacy* of the threat of extinction ("degree of threat")
- The *likelihood* that recovery efforts will succeed ("recovery potential")

These criteria are used, in fact, by the U.S. Fish and Wildlife Service in its current eighteen-point "recovery plan ranking system."[23]

So far, smooth sailing. But hard questions arise when you consider other possible ranking criteria. For example, what weight, if any, should be given to the following factors?

- Economic cost of recovery efforts
- Ecological value of the species (Is it, for instance, a "keystone species," crucial to the health of a particular ecosystem?)
- Scientific value/interest of the species
- Aesthetic value of the species
- Recreational value of the species
- Religious, symbolic, cultural, or historical value of the species
- Whether native species should be favored over invasive ones

---

22. Wilson, *The Diversity of Life*, p. 351. In a similar vein, Pope Francis writes: "Each year we see the disappearance of thousands of plant and animal species which we will never know, which our children will never see, because they have been lost for ever. . . . Because of us, thousands of species will no longer give glory to God by their very existence, nor convey their message to us. We have no such right." Pope Francis, *Laudato Si'*, §33.

23. U.S. Fish and Wildlife Service, "Policy and Guidelines for Planning and Coordinating Recovery of Endangered and Threatened Species," May 25, 1990, pp. 4-5. *USFWS National Digital Library* (online). Web. 20 July 2020.

- Whether "higher" (or more complex) species should be favored over "lower" (or less complex) ones
- Usefulness of the species to humans
- Harmfulness of the species to humans
- Popularity of the species (For example, should priority be given to "flagship" or "charismatic" species, such as pandas or elephants?)
- Level of public support for preservation and recovery efforts
- Level of expert support for preservation and recovery efforts

Readers must decide whether any or all of these factors should play a role in species preservation policy, and if so, how much relative weight they should be given. For instance, should the scientific value of a species take precedence over its aesthetic value? Or should charismatic species generally be prioritized over species that may possess greater ecological value? Such decisions are difficult and bound to be contentious. But any comprehensive species preservation and recovery plan will need to wrestle with such tough questions.

Chapter Summary

1. This chapter asks: How concerned should we be with species extinction and biodiversity loss? Though several mass extinctions have occurred before in earth history, the current extinction crisis—the so-called Sixth Extinction—is unprecedented and due almost entirely to human causes.
2. There are both anthropocentric (human-centered) and nonanthropocentric (non-human-centered) reasons for concern about biodiversity loss. Anthropocentric reasons include:
    - Many species are directly useful to humans.
    - Many species are indirectly useful to humans.
    - Some species, though not now useful to humans, might someday become useful.
    - Some species, though not of practical or economic value, may be valued for other reasons, such as their recreational, aesthetic, historical, or scientific importance.

3. Nonanthropocentric reasons for concern about biodiversity loss include:
   - Species extinction means the permanent loss of a biological "template" or "type." Such a loss is usually of far greater import than the loss of an individual exemplification (or "token") of that type.
   - Extinction of a single species can snowball into "extinction cascades" involving many other species, thus greatly multiplying ecological damage and biodiversity loss.
   - Biodiversity—richness and variety of life—has value in itself, inasmuch as it allows myriad forms of excellence, beauty, and complexity to flourish.
4. Unfortunately, it is not possible to save all species from extinction; biodiversity loss has now become so severe that we simply lack the human and financial resources to do so. Ranking principles must therefore be established. Various factors might be weighed in such a ranking system, including how close the species is to extinction; its role in maintaining ecosystem health; its economic, scientific, or cultural value; and how popular the species is with the general public. Exactly which priority principles should be adopted is a matter of considerable debate.

Discussion Questions

1. What anthropocentric reasons are there for preserving species and biodiversity? How compelling are those reasons?
2. What nonanthropocentric reasons are there for preserving species and biodiversity? How convincing are they?
3. Evaluate the following argument: We shouldn't be terribly concerned with species extinction, because nature always bounces back in the long run. In a few million years earth will have just as much biodiversity as it has today.
4. Should we attempt to save all the endangered species we can, regardless of cost or effort?
5. Are there some species whose extinction we should actually welcome—or even celebrate? If so, which?
6. Though many large North American mammals (e.g., elk, bison, grizzly bears, mountain lions, and wolves) are not currently endangered, their

numbers today are a tiny fraction of what they were before the European discovery of the New World. Should efforts be made to greatly increase such populations? If so, how could this safely be done?

7. What role, if any, should zoos play in conservation efforts?
8. When it's not possible for us to save all endangered species in a particular region or location (e.g., because of limited resources or loss of habitat), how should we prioritize which ones to save?

Further Reading

For an absorbing account from the frontlines of the fight against extinction, see Elizabeth Kolbert's *The Sixth Extinction: An Unnatural History* (New York: Henry Holt and Co., 2014). For the effect of climate change on species extinction, see Thomas E. Lovejoy and Lee Hannah, eds., *Biodiversity and Climate Change: Transforming the Biosphere* (New Haven, CT: Yale University Press, 2019). Bryan G. Norton, *Why Preserve Natural Variety?* (Princeton, NJ: Princeton University Press, 1987) offers a careful and sophisticated discussion of why biodiversity matters. Edward O. Wilson, *The Diversity of Life* (New York: W. W. Norton, 1992) is a modern classic and a full-throated defense of biodiversity preservation. For a lucid and compact discussion of reasons to preserve species by a leading environmental ethicist, see Holmes Rolston III, "Biodiversity," in Dale Jamieson, ed., *A Companion to Environmental Philosophy* (Malden, MA: Blackwell, 2001), pp. 402–15.

Chapter 13

Climate Change

The basic facts of climate change are no longer in serious dispute. Since roughly the middle of the eighteenth century, humans have been heating up the planet by injecting billions of tons of heat-trapping greenhouse gases into the atmosphere. Most of those gases come from burning fossil fuels, such as oil, natural gas, and coal. Each day, humans emit more than 140 million tons of carbon dioxide and other greenhouse gases. As a result, temperatures are rapidly rising around the globe. This causes killer heat waves, superstorms, wildfires, droughts, floods, rising seas, biodiversity loss, spreading tropical diseases, warming and acidifying oceans, and growing numbers of climate refugees. Since 1750, average global temperatures have increased by 1.1°C. If present trends continue, temperatures are projected to rise an additional 3–4°C by 2100. The last time the earth was that hot was fifty million years ago, when there were palm trees in Alaska and the oceans were hundreds of feet higher than they are today. At such temperatures, large portions of the globe would be uninhabitable, hundreds of major cities would be underwater, massive wildfires would burn out of control, crops would fail, global biodiversity would plummet, and damage from monster hurricanes, floods, and other climate-related disasters would cost trillions of dollars a year.[1]

Climate change is the mother of all environmental problems. It is also the mother of all ethical, political, economic, scientific, and technological problems. To stave off catastrophic climate change, carbon emissions must be quickly and drastically cut. Essentially, the entire global economy will need to be rapidly "decarbonized" and converted to renewable clean-energy sources such as wind, solar, geothermal, hydropower, and biomass. Such a shift will require profound changes in the economy, in international relations, and in how people work, travel, power their homes, and live their daily lives. Combatting climate change will be tremendously costly and will succeed only if formidable obstacles can be overcome. As leading climate ethicist Stephen Gardiner notes, climate change

1. David Wallace-Wells, *The Uninhabitable Earth: Life After Warming* (New York: Tim Duggan Books, 2019), pp. 12–15.

amounts to a kind of "perfect moral storm" because of the unprecedented cluster of difficult and often novel ethical, political, economic, and scientific challenges it presents.[2] Yet the battle against climate change may also present a unique opportunity to build whole new Green industries, create millions of high-paying new jobs, abandon unsustainable environmental practices, and make fundamental, long-overdue political and economic changes.[3]

13.1 The Urgent Challenges We Face

In the historic Paris Agreement (drafted in 2015, in force since 2016), over 190 nations agreed to work to keep the increase in average global temperatures well below 2°C above preindustrial levels. Even more ambitiously, the nations agreed to pursue efforts to limit the increase to 1.5°C, the point at with truly catastrophic climate effects will begin to occur.[4] To see how ambitious even this smaller increase is, consider what would be needed to achieve it.

According to a 2018 U.N. report, to keep warming below 1.5°C, global carbon emissions would need to be cut 45 percent below 2010 levels by 2030 and reach net-zero emissions by 2050.[5] This means that over the next decade global emissions must begin to fall rapidly (they rose 2.1 percent in 2018 and are projected to continue to rise until at least 2030, though the 2020 coronavirus pandemic may

---

2. Stephen M. Gardiner, *A Perfect Moral Storm: The Ethical Tragedy of Climate Change* (New York: Oxford University Press, 2011).

3. See Naomi Klein, *This Changes Everything: Capitalism vs. the Climate* (New York: Simon & Schuster, 2014).

4. Scientists warn that at two degrees of warming, there is a significantly greater risk of disastrous climate impacts resulting from various feedback loops, such as the sudden release of vast quantities of methane currently frozen in Arctic permafrost, or loss of the albedo effect as heat-reflecting ice is replaced by heat-absorbing dark water and soil, or the disruption of the Atlantic "conveyor belt" that brings warm waters from the equatorial South, thus preventing northern Europe from slipping into another Ice Age. For more on such potentially catastrophic feedback loops, see Wallace-Wells, *The Uninhabitable Earth*, pp. 22, 46.

5. The Intergovernmental Panel on Climate Change, Special Report, "Global Warming of 1.5°C," October 8, 2018. Web. 18 April 2020. The IPCC projections assume that in addition to drastic emissions cuts, significant amounts of carbon will need to be removed from the atmosphere by various "carbon capture and sequestration" strategies. Aside from planting trees, which would have to be done on a massive scale and would take some time to become effective, such carbon removal strategies and technologies are mostly just pipe dreams today.

slow this increase for a time).[6] If major emissions cuts are not achieved by 2030, the chances of limiting warming to 1.5 degrees are extremely bleak.

To appreciate how difficult it is to muster the political will to limit the impacts of dangerous climate change, consider some of the climate policies proposed by Washington state governor Jay Inslee, a Democratic candidate for U.S. president in 2019. Inslee's proposals, contained in six reports totally more than two hundred pages, call for:

- A $9 trillion-dollar spending package paid for in part by a "climate polluter fee" on the fossil fuel industry.
- A 50 percent cut in U.S. carbon emissions by 2030, and net-zero emissions by 2045.
- By 2030, all new passenger vehicles and medium-duty trucks and buses sold in the United States must be fully electric, zero-emission vehicles. To encourage a quick transition to electric vehicles, generous trade-in rebates would be offered through a "Clean Cars for Clunkers" program.
- A kind of "G.I. Bill" would provide aid to the millions of workers who would lose their jobs in the clean-energy transition. A similar "Re-Power Fund" would assist communities especially hard-hit by job losses.
- A carbon tax would be imposed to reduce demand for gasoline and other polluting energy sources, and fossil fuel subsidies would be ended.
- National commitments would be made to achieve 100 percent clean electricity generation by 2035 and 100 percent clean new buildings by 2030, along with significant spending to retrofit existing buildings to reduce energy demand and fossil fuel usage.[7]

As wildly ambitious as all that may sound, it greatly understates the cost and scale of a conversion to a net-zero-emissions society by midcentury. Today,

---

6. Due to widespread coronavirus-related lockdowns in the first few months of 2020, global emissions plummeted by around 17 percent in April 2020. However, as countries reopened over the next several months, emissions quickly rose to close to previous levels. Scientists say that the short-term reductions—if indeed they prove to be short-term—are far too little to significantly slow the pace of climate change. Emma Newberg, "Carbon Emissions Sharply Rebound as Countries Lift Coronavirus Restrictions," *CNBC*, June 18, 2020 (online). Web. 21 July 2020.

7. Marianne Lavelle, "Jay Inslee on Climate Change: Where the Candidate Stands," *Inside Climate News*, August 21, 2019. Web. 19 April 2020. Most of the other 2020 Democratic presidential candidates adopted climate change plans similar to Inslee's, including eventual Democratic presidential nominee Joe Biden. In July 2020, Biden announced an updated climate plan that largely mirrors Inslee's, but has a much cheaper price tag ($2 trillion versus Inslee's $9 trillion). Alana Wise, "Biden Announces $2 Trillion Climate Plan," *NPR*, July 14, 2020 (online). Web. 21 July 2020.

the United States contributes less than 15 percent of the world's carbon emissions. Any drastic cuts in U.S. emissions would largely be futile, therefore, unless other major polluting nations (especially China, India, and the Russian Federation) follow through with similar actions. In recent decades, China (which is now by far the world's biggest carbon-polluter, accounting for 28 percent of global emissions) has built hundreds of coal-fired energy plants and is unlikely to decommission those any time soon. Moreover, many poor countries can't afford to switch over to clean energy without massive infusions of aid from affluent countries like Germany and the United States. Nor does Inslee account for the huge costs that would be entailed by replacing gas-guzzling airplanes, ships, diesel rail engines, farm machinery, furnaces, stoves, hot-water heaters, lawnmowers—not to mention military aircraft, tanks, armored personnel carriers, and other such vehicles. (The U.S. military alone emits more greenhouse gas emissions annually than over 140 countries.[8])

Aside from such issues of cost and implementation, there are huge political obstacles to any rapid conversion to clean energy, particularly in the United States. Inslee's proposals would require bold, decisive political action and a World War II–like mobilization of resources, sustained over a period of many decades. The gridlock of Washington politics would need to be overcome, the power of special interests curbed, and somehow American voters (large numbers of whom are climate skeptics[9] or unconcerned about climate change because they believe the end of the world is near[10]) would need to be convinced to support policies that result in higher taxes, the bankruptcy of whole industries, massive economic dislocations, and higher gas prices. For all these reasons, it is very difficult to believe that the United States—or many of the other major carbon-polluting nations—will be able to achieve anything close to net-zero emissions by midcentury. It seems certain, therefore, that catastrophic and enormously costly climate change will occur.

Perhaps surprisingly, there are climate optimists. Some believe that game-changing new technologies—a so-called **techno-fix**—can be found to

---

8. Niall McCarthy, "Report: The U.S. Military Emits More CO2 than Many Industrialized Nations," *Forbes*, June 13, 2019. Web. 22 April 2020.

9. According to one recent poll, only 71 percent of Americans believe that human activity contributes to climate change. Jennifer de Pinto et al., "Most Americans Say Climate Change Should be Addressed Now—CBS Poll," *CBS News*, September 15, 2019. Web. 22 April 2020.

10. In a 2016 poll, 26 percent of American evangelicals and born-again Christians said we don't have to worry about climate change because the End Times are at hand. Connie Roser-Renouf et al, "Global Warming, God, and 'End Times,'" *Yale Program on Climate Change Communication*, July 26, 2016. Web. 24 April 2020.

mitigate global warming. This is an idea that is slowly gaining traction. Current techno-fix proposals include fertilizing the oceans with iron to spur phytoplankton growth, planting billions of carbon-storing trees, building vast plantations of carbon dioxide removal machines, capturing emissions from power plants and burying them deep underground, and deploying huge mirrors in space to reflect incoming sunlight. Each of these modes of "geoengineering" has significant drawbacks. At present, the most often discussed techno-fix option is to inject sulfur into the upper atmosphere, thereby cooling the earth by reflecting small amounts of sunlight back into space. This could be done at relatively low cost and with existing technology (for example, by aerosol spraying from high-altitude balloons or specially modified planes). A further advantage is that sulfur injection is something that sometimes occurs naturally. The explosion of Mount Pinatubo in 1991 pumped twenty million tons of sulfur dioxide into the atmosphere, causing average global temperatures to drop by at least 0.5°C over a period of three years—a significant (but short-term) reduction. A further advantage is that sulfur injection, unlike some other geoengineering proposals, has environmental impacts that are relatively well understood. Critics object that sulfur injection would "bleach" the sky (turning it whitish rather than blue), might have adverse effects on the ozone layer, could cause droughts or hurricanes in some parts of the planet, and could easily inflame global tensions as some parts of the world are benefited by sudden global cooling and some are harmed. They also note that, once deployed, sulfur injection could lead to dangerously rapid warming if the sprayings were ever halted while global CO2 levels are still high. Critics also point out that sulfur injection would do nothing to address ocean acidification and other climate problems that result simply from high levels of carbon and other greenhouse gases in the atmosphere. Finally, many critics find the whole idea of so-called "solar radiation management" reckless and arrogant, and fear that it will become an excuse for major carbon-polluting nations to continue dragging their feet on emissions reduction, thus greatly increasing the risk of climate disaster.[11]

Despite these concerns, a strong case can be made that serious private- and government-funded research on sulfur injection and other forms of geoengineering should begin immediately.[12] While everyone hopes that such hazardous, costly, and extreme steps to save the planet will never be needed, the stakes are simply too high to ignore the risks.

---

11. For a detailed discussion of the pros and cons of geoengineering, see Gardiner, *A Perfect Moral Storm*, pp. 339–96.

12. For an extended argument along these lines, see David Keith, *The Case for Geoengineering* (Cambridge, MA: MIT Press, 2013).

MAKING A DIFFERENCE

Greta Thunberg: A Little Child Will Lead Them

Swedish climate activist Greta Thunberg is seventeen, but the waif-like girl with pigtails looks much younger. In August 2018, when she fifteen, Thunberg was inspired by the school strikes that followed the Parkland, Florida, mass shooting in February of that year. She urged her school friends to join her at a school climate strike, but her classmates showed little interest. So Greta decided to do it on her own. She painted a homemade sign that read *Skolstrejk för klimatet* (school strike for climate) and parked herself in front of the Swedish parliament in Stockholm, demanding that officials get serious about fighting climate change. Soon other schoolchildren joined her. Media outlets picked up on the story. In the months that followed, tens of thousands of school-strikers in cities around the world joined her cause and a climate icon was born.

Since then, Thunberg has participated in dozens of climate protests and given high-profile speeches at venues throughout the world, including the British, French, and European parliaments, and at the U.N. Climate Action Summit. In January 2019, at the World Economic Forum in Davos, Switzerland, Thunberg gave her famous "Our House Is on Fire" speech, calling on world leaders to take bold, immediate action to address the climate crisis. Refusing to fly because of air travel's large carbon footprint, Thunberg twice crossed the Atlantic on sailing vessels to attend climate gatherings. In December 2019 she was named *Time* magazine's Person of the Year.

Thunberg may seem an unlikely hero. Diagnosed with Asperger's syndrome and an anxiety-related speech disorder, Thunberg was bullied in school and had a very difficult childhood. Yet Thunberg makes light of her troubles. "If I were like everyone else," she explains, "I wouldn't be able to sit for hours and read things I'm interested in."[13]

Not surprisingly, Thunberg has her critics, including U.S. President Donald Trump and Brazilian president Jair Bolsonaro, who called her "a

13. Quoted in Charlotte Alter, Suyin Haynes, and Justin Worland, "Greta Thunberg: *Time* 2019 Person of the Year," *Time* magazine, December 23, 2019 (online). Web. 21 July 2020.

brat." When asked by an interviewer what she thought of the haters, she replied, "I think it is quite hilarious when the only thing people can do is mock you, or talk about your appearance or personality, as it means they have no argument or nothing else to say."[14]

While world leaders have dithered and global emissions have continued their dangerous and inexorable rise, Thunberg seems to have finally galvanized a reaction. Public opinion polls show record levels of environmental concern, investments in clean energy are sharply up, sales of children's books dealing with climate change are soaring, "flight shame" is prompting many to reconsider air travel, Green political parties are growing, and politicians around the world are beginning to talk seriously about cutting emissions. "You have woken us up," British labor politician Ed Miliband said to Thunberg after her speech to the British Parliament. Climate activists, finally seeing the needle beginning to move, speak of "the Greta effect." Hopefully, the world will wake up.

13.2 Climate Ethics

Climate change presents huge and unprecedented ethical challenges. As Dale Jamieson has pointed out, climate change creates ethical issues that leave ordinary morality "flummoxed, silent, or incorrect."[15] Human ethics evolved mostly to deal with discrete, easily identifiable harms to specific individuals within relatively small, close-knit societies. Climate change involves harms to anonymous individuals, most of them far distant from us in space and time, caused by such simple and seemingly innocuous acts as turning on a light or driving to the grocery store. Moreover, as climate ethicist Stephen M. Gardiner has noted, existing ethical theories seem ill equipped to sort through the dilemmas of climate ethics; we simply lack robust theory in key areas such as intergenerational ethics, environmental philosophy, international justice, and scientific uncertainty.[16] All of this may help explain why so many people find it difficult to see climate change as the onrushing planetary train wreck that it is.

---

14. Quoted in Suyin Haynes, "'Now I Am Speaking to the Whole World': How Teen Climate Activist Greta Thunberg Got Everyone to Listen," *Time* magazine, May 16, 2019 (online). Web. 21 July 2020.

15. Dale Jamieson, *Reason in a Dark Time: Why the Struggle Against Climate Change Failed—and What It Means for Our Future* (New York: Oxford University Press, 2014), p. 147.

16. Gardiner, *A Perfect Moral Storm*, p. 399.

In fact, though, the ethical challenges of climate change are unavoidable, both for individuals and for governments. The climate crisis forces all of us to ask hard questions about our respective carbon footprints. Are there ways I could eat, travel, commute, shop, exercise, and heat and cool my home that would significantly lower my greenhouse emissions? How important is it to vote for political candidates who support strong climate action? Is it ethical or wise to have children when they, too, will surely contribute to climate change and may well face a bleak future on a sick or dying planet? Governments also must confront urgent issues of climate ethics where many of the usual norms of international politics no longer apply. The normal assumption of international politics—that each nation should chiefly concern itself with its own relatively short-term national interest—is a certain recipe for climate disaster. As Gardiner notes, international climate politics constitutes a classic "tragedy of the commons" scenario in which each country has an incentive to free ride on the actions of others but where doing so leads to inevitable tragedy for all.[17] The only solution would seem to be to discourage free riding through some system of enforceable sanctions or other means of coercion. Yet this would require a degree of international agreement and cooperation that many today find hard to imagine.

Assuming that an effective system of international cooperation can be worked out, and worked out quickly, many crucial ethical issues remain. Let's briefly examine two: (1) Who should bear most of the costs of mitigating and adapting to climate change? (2) As countries transition to clean energy, what would be a fair way of allocating further emissions?

It is widely agreed among climate ethicists that the costs of fighting climate change should be borne mainly by rich industrialized countries. Several reasons are commonly offered for this. First, rich countries are responsible for most of the world's carbon pollution (80 percent of total global emissions come from just ten countries). Simple schoolyard maxims like "You broke it, you fix it" and "Clean up your own mess" would thus suggest that they pay the lion's share of the costs. Second, rich countries have reaped most of the rewards of cheap carbon energy. In large measure, they have become rich by harnessing the power of cheap energy. Third, poor nations are expected to suffer disproportionately from the ravages of climate change, such as flooded cities, scorched croplands, and damage from superstorms. Finally, most poor nations cannot afford large expenditures to combat and adapt to climate change. Most lack the resources to do things like build expensive seawalls, air-condition homes

17. Gardiner, *A Perfect Moral Storm*, p. 28.

and businesses to protect against deadly heat waves, and construct whole new clean-energy grids.

Such arguments seem sound, but of course any massive transfer of wealth from rich to poor nations to fight climate change is probably unrealistic, at least any time soon. Few affluent nations would be willing to shell out vast sums of climate aid to poor countries when they themselves face will huge bills to pay for climate mitigation and adaptation within their own borders. Certainly in the United States any politician who proposed spending billions of taxpayer dollars to build wind farms in Borneo or seawalls in Bangladesh would be quickly drummed out of office. On a more principled level, politicians in affluent nations must be mindful that the fight against climate change cannot be won without the financial and human resources that strong economies make possible. Ultimately, direct foreign assistance and wealth transfers may prove less helpful in combating climate change than innovative technological breakthroughs emerging from the labs and research centers of rich industrialized nations. That said, a compelling case can be made that wealthy nations do have a moral duty to provide generous assistance—financial and otherwise—to poor nations struggling to deal with climate change. Which countries should pay, and how much, should probably be worked out by binding international agreement. A fair funding formula would likely factor in both past emissions and ability to pay.

A second critical ethical issue facing governments is to determine how future global carbon emissions should be allocated. If global emissions are to be cut by a hefty 40–50 percent by 2030, as climate scientists urge, quotas will need to be adopted and enforced to deter free riders and ensure basic fairness. Several proposals along these lines have been offered. One is equal per capita emissions, perhaps combined with a trading system that would permit countries that wish to exceed their quota to purchase emissions credits from low-emitting countries.[18] Equal per capita emissions has the appearance of fairness, but it ignores the fact that some countries have much higher energy needs than others. It also has no chance of gaining acceptance, since it would cripple the economies of high-polluting countries like the United States, the Russian Federation, and Canada.

---

18. For defenses of equal per capita emissions, see Peter Singer, *One World: The Ethics of Globalization* (New Haven, CT: Yale University Press, 2002), pp. 39–40; and Dale Jamieson, "Climate Change and Global Environmental Justice," in Paul Edwards and Clark Miller, eds., *Changing the Atmosphere: Expert Knowledge and Global Environmental Governance* (Cambridge, MA: MIT Press, 2001), pp. 287–307.

A better proposal is to impose equal burdens on all countries.[19] All nations, for example, might be required to slash their emissions by 30 percent by 2030. This, too, has the appearance of fairness, but in fact it would greatly advantage wealthy countries, since they are better positioned to absorb cuts in their quality of living and have the resources to rapidly convert to cleaner forms of energy. In practice, an equal-burdens approach would probably lock in poverty for many developing countries for decades to come.

A third proposal is to recognize a right to "subsistence emissions," that is, emissions that are necessary for survival and some minimal quality of life.[20] This seems reasonable and accords with widely accepted international norms guaranteeing basic human rights.[21] There are obvious problems spelling out what should count as a "subsistence emission." Notions of what constitutes a "minimal quality of life" vary widely around the globe. But a bigger problem is that it says nothing about how non-subsistence emissions should be allocated, and thus is only a small step in the direction of finding a fair allocation formula.

At present, all academic discussions of emissions allocations have an air of unreality because the unmovable stone of climate politics is that no wealthy nation would accept any plan that threatens serious economic decline, political instability, or requires massive transfers of taxpayer-funded wealth to poor nations. A more realistic scenario would involve scaling up the climate plan that formed the basis of the Paris Agreement (2016). Such a strategy would require much more ambitious "nationally determined contributions" and probably enforceable sanctions against countries that opt out or fall significantly short of their emissions-reduction pledges. While the fight against climate change has been agonizingly and inexcusably slow, it is bound to pick up steam as the impacts of climate change become increasingly more obvious and severe. Perhaps, as Naomi Klein has argued, climate change will "change everything."

Klein believes that because of climate change, "our economic system and our planetary system are now at war."[22] In her view, we can no longer afford to live, spend, and consume as we have before. Now, she argues, the economies of major carbon-polluting nations must be heavily regulated to ban or discourage

---

19. See Martino Traxler, "Fair Chore Division for Climate Change," *Social Theory and Practice* 28 (2002), pp. 101–34.

20. See Henry Shue, "Subsistence Emissions and Luxury Emissions," *Law and Policy* 15:1 (1993), pp. 39–59.

21. For example, Article 25 of the UN Universal Declaration of Human Rights (1948) declares that "everyone has the right to a standard of living adequate for the health and well-being of himself and of his family, including food, clothing, housing and medical care and necessary social services."

22. Klein, *This Changes Everything*, p. 21.

all manner of planet-warming activities. Klein also believes that our current politics are at war with our planet. When wealthy corporations and powerful special interests control the levers of government, no effective response to climate change is possible. Only if citizens insist on genuine self-government and an economic system that puts people—and the planet—ahead of profits can the climate fight be won.[23] For these reasons, Klein believes that the climate crisis presents an unparalleled opportunity for fundamental economic and political change. She is less clear, however, on just what form those big changes should take.

Without question, climate change will profoundly change all our lives. Though governments and large organizations probably will have to do most of the heavy lifting in fighting climate change, individuals must also do their part. Some simple things many of us can do to reduce our individual carbon footprints are: fly and drive less, use public transportation or bike to work, reduce food waste, unplug items not in use, eat less beef, consider installing solar panels on your home, switch to the renewable energy option from your energy provider, lower the temperature in your water heater, wash your clothes in cold water and line-dry them, replace natural-gas appliances with clean-energy electric ones, choose low-carbon recreational activities (e.g., hiking or cycling) over high-carbon ones (e.g., power-boating or golf), recycle, reduce unnecessary purchases, eat more local foods, and shop and invest in earth-friendly ways. If we each do our part, our planet can still have a Green tomorrow.

Chapter Summary

1. Climate change is the mother of all environmental problems; it is also the mother of all ethical, political, economic, scientific, and technological problems. Climate ethicist Stephen M. Gardiner has called climate change a "perfect moral storm," because it poses a host of unprecedented ethical challenges with no easy solutions.
2. In the landmark 2016 Paris Agreement, signatory nations committed themselves to efforts to keep global warming to 1.5°C above preindustrial levels. To have any reasonable chance of achieving this, drastic cuts in carbon emissions must begin immediately and net-zero emissions must

23. Klein, *This Changes Everything*, p. 364.

be achieved by 2050. This will require the rapid "decarbonization" of the global economy and a swift transition to clean energy sources such as wind and solar. This, in turn, would require substantial public expenditures, profound economic and political changes, and major alterations in our daily lifestyles. Given the many obstacles to such rapid, transformative change, many climate experts are skeptical that the necessary cuts can be made. It seems likely, therefore, that catastrophic climate change will occur, resulting in rising seas, flooded cities, significant biodiversity loss, damaging superstorms, droughts, raging wildfires, dying oceans, crop losses, large numbers of climate refugees, and deadly heat waves.

3. There are, however, climate optimists. Many believe that some game-changing "techno-fix" will be found to climate change. Though all of the most-discussed technological solutions have major drawbacks, prudence dictates that we begin serious research on solar radiation management, carbon removal, and other possible techno-fixes immediately. The consequences of doing otherwise could be dire.
4. Climate change poses unique and extremely difficult ethical challenges, both for individuals and for governments. The climate crisis requires us to reflect on possible but uncertain and often unwitting harms to strangers, future generations, and nonhuman organisms, many of whom are remote from us in space or time. According to many climate ethicists, evolution has not equipped humans to grapple well with such unusual ethical dilemmas. Current ethical theories may not be of great help either, because we currently lack well-developed theories in relevant fields such as intergenerational ethics, international justice, and interspecies ethics.
5. One key question of climate ethics is: Who should bear the costs of fighting climate change? Most climate ethicists agree that affluent industrialized nations should pay the lion's share of the costs, but there are major obstacles to making this a reality.
6. Another important question of climate ethics is: How should future carbon emissions be allocated? Various proposals have been offered, but all are problematic. Though far from perfect, a framework of "nationally determined contributions," similar to that adopted in the 2016 Paris Agreement, might be the most viable option. To be effective, however, this would require enforcement mechanisms that may be difficult to achieve.

7. In her much-discussed book, *This Changes Everything: Capitalism vs. the Climate* (2014), Naomi Klein argues that no solution to climate change is possible without changing our capitalist economic system and our broken, special-interests-dominated politics. She believes that the climate crisis offers an unparalleled opportunity to fundamentally rethink our failed economic and political systems and change them for the better.

Discussion Questions

1. How concerned should we be about climate change?
2. How optimistic are you that a "techno-fix" will be found to the climate crisis?
3. What do you think of Governor Inslee's climate plan? Is it realistic?
4. What does Stephen Gardiner mean when he calls climate change the "perfect moral storm"? Do you agree?
5. Do you agree that that affluent, industrialized nations should pay the bulk of the costs to address climate change?
6. What's the fairest and most practical way to allocate future global carbon emissions?
7. Do you agree with Naomi Klein that our political and economic systems are now at war with our planetary system? If so, what's the best solution?

Further Reading

For a harrowing account of the grave dangers climate change poses, see David Wallace-Wells, *The Uninhabitable Earth: Life After Warming* (New York: Tim Duggan Books, 2019). For an excellent collection of readings on the ethical challenges posed by climate change, see Stephen M. Gardiner, Simon Caney, Dale Jamieson, and Henry Shue, eds., *Climate Ethics: Essential Readings* (New York: Oxford University Press, 2010). Stephen Gardiner's Introduction to this volume, titled "Ethics and Global Climate

Change," offers a clear and comprehensive survey of the main issues in climate ethics. Gardiner's monograph, *A Perfect Moral Storm: The Ethical Tragedy of Climate Change* (New York: Oxford University Press, 2011), is a landmark in the field. Most of that work is descriptive, seeking to explain why climate change poses such a deep moral challenge. But Gardiner also proposes some helpful principles of a "transitional ethics" as we await development of more robust theories of climate ethics. For an interesting debate on Gardiner's perfect-moral-storm view, see Stephen M. Gardiner and David A. Weisbach, *Debating Climate Ethics* (New York: Oxford University Press, 2016). For a thought-provoking treatment of individual ethical responsibilities in a warming world, see Walter Sinnott-Armstrong, "It's Not My Fault: Global Warming and Individual Moral Obligations," in Richard Sinnott-Armstrong and Richard Howath, eds., *Perspectives on Climate Change* (Amsterdam: Elsevier, 2005), pp. 221–53; reprinted in Gardiner et al., eds., *Climate Ethics*, pp. 332–46. For a thoughtful critical response to Naomi Klein's book, *This Changes Everything*, see Elizabeth Kolbert, "Can Climate Change Cure Capitalism?" *The New Yorker*, December 4, 2014, available online.

Chapter 14
Environmental Disobedience

Many people who care about the environment are not content simply to talk about saving the planet; they believe that we must *act*. Such people may choose, for example, to ride a bike to work or to install solar panels on their home. Some may even become environmental activists, organizing eco-literacy workshops, participating in pro-environmental rallies, and lobbying their political representatives to support Green causes. These are all legal and widely admired forms of ecological action. But what about more extreme forms of environmental activism? Is it ever ethically permissible, for example, to break the law in pursuit of an environmental goal? Might it sometimes be morally justifiable, for example, to practice peaceful, nonviolent civil disobedience to protest or prevent some serious harm to the environment? More questionably, might it sometimes be ethical (or "necessary") to engage in violent criminal acts—acts of what are sometimes called "eco-terrorism" or **ecosabotage**—to advance some urgent or high-priority environmental cause? In this chapter we'll examine these complex and important issues.

14.1 Varieties of Environmental Protest

Ecological protests can take many forms, both legal and illegal. In liberal democracies, many kinds of protest are permitted by law. Marches, rallies, boycotts, petitions, speeches, signs, letter-writing campaigns, and phone calls to political representatives are all perfectly legal in democratic (and many nondemocratic) societies. Depending on the circumstances, such protests may be unwise, imprudent, or unethical, but they are not illegal.

Illegal protests can also take a variety of forms. One is **conscientious refusal**, when a protester publicly refuses to obey a law or legal command for personal ethical or religious reasons. Henry David Thoreau's refusal to pay a poll tax because he believed that the Mexican War (1846–1848) was imperialistic and unjust is a well-known example of conscientious refusal.

Another form of illegal protest is **conscientious evasion.** This occurs when a protester refuses to obey a law for personal ethical or religious reasons but does not wish to get caught and tries to keep his or her actions secret. Northern abolitionists who, prior to the American Civil War, covertly helped slaves escape to freedom on the Underground Railroad were practicing conscientious evasion.

A third kind of unlawful protest is **militant action.** This is a type of protest or direct action that involves coercion, property damage, or violence. Ecological activists who spike trees, sabotage bulldozers, spray-paint buildings, break windows, ram whaling ships, burn down lumber companies, or torch SUV dealerships are engaged in militant action.

A fourth form of illegal protest is **civil disobedience.** The term "civil disobedience" is sometimes used broadly to describe any kind of conscientious noncompliance with law, but it is useful to define it more narrowly, as a public, nonviolent, deliberately unlawful act of protest, usually done with the aim of bringing about a change in the law or the policies of the government.[1] This is the type of illegal protest practiced by Martin Luther King Jr. and his supporters during the civil rights era, and it is widely employed in environmental protests today.

Civil disobedience differs in important ways from other forms of unlawful protest. Unlike militant action, civil disobedience does not involve force or violence. And unlike conscientious evasion, civil disobedience is always a public act intended to draw attention to what is believed to be an intolerable evil, harm, or injustice.

14.2 The Justifiability of Unlawful Environmental Protest

Some political theorists argue that in a democratic society it is never morally justifiable to break a law.[2] On reflection, however, few people would accept such

1. For similar definitions, see Hugo A. Bedau, "On Civil Disobedience," *Journal of Philosophy* 58 (1961), pp. 653–61; and John Rawls, *A Theory of Justice* (Cambridge, MA: Harvard University Press, 1971), p. 364. Some theorists question whether civil disobedience must, by definition, always be peaceful or nonviolent. It is also debated whether civil disobedience must be "a political act" in the further sense of being based on widely shared principles of justice or other ethical or political values.

2. See, for example, T. H. Green, *Lectures on the Principles of Political Obligation* (London: Longmans, 1907), p. 111.

an absolutist view. For example, nearly everyone would agree that it is ethical for a motorist to break the speed limit if that is the only way to save the life of a critically injured child.[3] And most Americans would agree that Martin Luther King Jr.'s acts of peaceful, nonviolent civil disobedience were legitimate ways to protest blatant acts of racial discrimination.

Such cases make clear that the obligation to obey the law is not an absolute duty but at best a presumptive one. No doubt it is true, as Carl Cohen argues, that "[c]ivilized human life requires . . . that at least the vast majority of citizens recognize the authority of some law-making body, and that they accept the laws enacted by that body as their own, properly governing them."[4] But societies can be stable, just, and well-ordered even if some laws are intentionally violated. In fact, as the American civil rights movement makes clear, some acts of illegal disobedience may *contribute* to social justice and respect for the law over the long run. Assuming, therefore, that it is sometimes morally justifiable to break a law, we must ask whether lawbreaking is ever acceptable as a form of environmental protest, and if so, under what circumstances.

The issue is complicated by disagreements over the importance of environmental protection. Many groups that practice ecological civil disobedience or ecosabotage (also known as "ecotage" or "monkeywrenching") are motivated by what many see as extremist environmentalist views. For example, Dave Foreman, one of the co-founders of the radical environmental group Earth First!, believed that all forms of life have equal intrinsic value, rejected modern industrial civilization, and held that most parts of the earth should be allowed to return to wilderness.[5] With such views, it's not difficult to see why people like Foreman might be motivated to become militant "eco-warriors." But what about people with less radical views? Would a typical member of, say, the Sierra Club or some other mainstream environmental group, ever be justified

---

3. Some legal systems have a "necessity defense," which allows people to break a criminal law if doing so is necessary to avoid some greater harm. In such systems, speeding to save a life would not be illegal.

4. Carl Cohen, *Civil Disobedience: Conscience, Tactics, and the Law* (New York: Columbia University Press, 1971), p. 2.

5. Dave Foreman, *Confessions of an Eco-Warrior* (New York: Harmony Books, 1991), pp. 26–35. (Foreman, who remains active in the environmental movement, was eventually arrested, left Earth First!, and may hold somewhat more moderate views now.) In a similar vein, Ned Hettinger argues that even in democratic societies environmental activists have little reason to respect or obey the law because modern democracies are thoroughly anthropocentric, and thus gravely unjust to nonhuman nature. Ned Hettinger, "Environmental Disobedience," in Dale Jamieson, ed., *A Companion to Environmental Philosophy* (Malden, MA: Blackwell, 2001), pp. 505–8.

in practicing environmental civil disobedience or militant action? That's the central question addressed in this chapter.

For starters, we must note that acts of civil disobedience are generally easier to justify than are acts of ecosabotage or other forms of militant action. Acts of civil disobedience are public, peaceful, nonviolent, and demonstrate a kind of respect for the law in virtue of the protesters' willingness to go to jail (or face other legal repercussions) for their beliefs. Militant acts, by contrast, involve force or violence; are usually covert; often create serious risks to innocent third parties or to the militants themselves; frequently create a harmful backlash; may stoke fear in a manner analogous to terrorism;[6] manifest contempt for law and democratic processes; and may result in significant prison time, with all its attendant hardships and public expense. Moreover, because of their covert nature, groups that practice ecosabotage or other forms of environmental militancy often have problems coordinating their activities and can easily attract dangerous and unpredictable recruits to their causes. For these reasons, I believe that militant environmental action is nearly always morally indefensible. As Michael L. Martin argues, while we can imagine extreme cases where force or violence would be justifiable to prevent some grave environmental loss, advocates of environmental militancy face a heavy burden of proof to demonstrate the acceptability of the criminal and often violent acts they support.[7] Accordingly, for the remainder of this chapter, I will set aside the issue of militant action and focus exclusively on the moral justifiability of civil disobedience as a form of environmental protest. Our main question, therefore, will be whether nonviolent, illegal sit-ins, occupations, blockades, trespasses, and other such acts of civil disobedience can ever be a legitimate means of protesting or seeking to prevent environmental harms.

Since the 1960s, the morality of civil disobedience has been extensively debated. Some theorists have examined the issue from a purely consequentialist standpoint (focusing exclusively on the good or bad consequences that acts

6. There is considerable debate over whether acts like tree spiking, monkeywrenching, and other such environmentally motivated militant crimes against property are properly classified as acts of "terrorism" (specifically, "eco-terrorism"). Especially after the 9/11 attacks, the FBI often treated them as such, but the motivations of self-described "eco-warriors" are usually quite different from those who use terror tactics to kill innocent civilians, intimidate populations, or coerce or overthrow governments.

7. Michael Martin, "Ecosabotage and Civil Disobedience," *Environmental Ethics* 12:4 (1990), pp. 291–310. Suppose you discover, for example, that a crazed, glue-sniffing crop duster is about to spray herbicide over a stand of old-growth redwoods. If there isn't time to alert the authorities, a militant act, such as disabling the airplane, would seem to be morally justified.

of civil disobedience may produce).[8] Others have looked at the matter from a purely duty-centered ethical perspective (focusing, for example, on competing moral principles and prima facie moral duties that bear on acts of civil disobedience).[9] And still others have looked at the issue from a mixed ethical perspective, focusing on consequences as well as on relevant ethical principles or virtues.[10]

From a consequentialist perspective, the moral justifiability of civil disobedience will depend largely on the relevant circumstances. Philosopher Carl Cohen has helpfully identified some of the pertinent consequentialist issues that must be weighed in the balance. These include:

- How serious is the harm or injustice whose remedy is being sought? Is it a clear and substantial evil, or one that is minor or more debatable?
- How pressing is the need for that remedy?
- Are the protestor's goals widely shared by the community, or will they be viewed as fringe or extremist?
- Have normal legal avenues for addressing the harm been actively pursued?
- How grave is the crime the protestor plans to commit?
- Is violence likely or threatened by the disobedient act?
- How great is the expense and inconvenience likely to be caused by the disobedience?
- Will a bad example be set?
- Will the act decrease respect for law in the community?
- Will there be a significant backlash against the protester and their cause?
- Is the disobedient act likely to succeed, or will it be a futile and antagonizing gesture?[11]

8. See, for example, Cohen, *Civil Disobedience*, pp. 120–28.

9. See, for example, Joel Feinberg, "Civil Disobedience in the Modern World," *Humanities in Society* 2:1 (Winter 1979), pp. 37–60; reprinted in Joel Feinberg and Hyman Gross, eds., *Philosophy of Law*, 4th ed. (Belmont, CA: Wadsworth, 1991), pp. 119–32.

10. See, for example, Hettinger, "Environmental Disobedience," pp. 498–509; and Lawrence B. Solum, "Virtue as the End of the Law: An Aretaic Theory of Legislation," *Jurisprudence* 9:1 (2018), pp. 6-18 (online). Web. 21 July 2020.

11. Cohen, *Civil Disobedience*, pp. 124–28. (I have slightly altered or paraphrased some of Cohen's queries.)

Sometimes these consequentialist questions will yield clear-cut answers, but often they will not. "In practice," as Carl Cohen notes, "the entering variables are so many, so complex, so difficult to measure, and so extended over time and place, as to render a clear resolution of the issues often impossible."[12] In general, however, consequentialist arguments for acts of environmental civil disobedience will fail, because the goals of the protesters are usually not widely shared by the larger community. Often, environmental protesters hold radical biocentric views that are seen as extreme by most citizens. As a result, many such protests are likely to be ineffective and stir a strong backlash.

Looking at probable consequences is one way of reflecting on the ethics of environmental civil disobedience. Another is to take a duty-centered (or "deontological") approach and appeal to widely shared moral principles that speak to issues of social responsibility and ecological conscience. What sorts of principles? Many might come into play, but some likely suspects might include:

- Promote just institutions and work toward their establishment, maintenance, and improvement (Principle of Just Institutions).
- Don't be a free rider; those who enjoy the benefits of society should also, in fairness, be prepared to bear its just burdens (Principle of Fair Play).[13]
- Obey the law unless you have a compelling reason not to (Principle of Legal Obedience).
- Don't cause unnecessary pain, suffering, or annoyance to others (Principle of Nonmaleficence).
- Do good to others (Principle of Beneficence).
- Care for the earth and its diverse community of life (The Environmental Principle).
- Be mindful of the needs of future generations (Principle of Future Generations).

These are broad principles that, at best, offer some relevant framing principles on the morality of particular acts of ecological civil disobedience. Somewhat more specific normative guidance might be gleaned from major international environmental documents such as the Earth Charter (2000).[14]

---

12. Cohen, *Civil Disobedience*, p. 124.

13. Difficulties arise if this is treated as more than a rule of thumb. For critical discussion, see Richard Dagger and David Lefkowitz, "Political Obligation," *Stanford Encyclopedia of Philosophy*, §4.3 (online). Web. 26 July 2020.

14. Available online at earthcharter.org. Web. 12 May 2020.

One can also, of course, appeal to personal values that may not enjoy wide public support. This is what Earth First! co-founder Dave Foreman does in his defense of environmental disobedience. Foreman argues that ecosabotage and other militant acts in defense of the environment are justified, in part, because all living things on earth have the same intrinsic value and the same right to live.[15] Here Foreman's argument runs into the same problems that so-called higher-law justifications of civil disobedience encounter. Many practitioners of civil disobedience, including Martin Luther King Jr., have sought to justify it by appealing to some law higher and more imperative than any human-made law or code of ethics (for example, "God's law" or "the law of nature"). The basic problem with this strategy, as Carl Cohen notes, is that it "appears impossible to reach any objective or reliable judgment about what the higher laws command or forbid (if there be any higher laws at all)."[16] For centuries, all kinds of zany or reactionary views have been supported by appeals to "higher law."[17] Usually no satisfactory rational grounds can be given for resolving disputes about what such a law demands. Consequently, there is a high burden of proof on any civil disobedient who resorts to higher-law appeals in defense of their unlawful conduct.

The upshot is that it is more difficult to justify civil disobedience as a means of ecological protest than is sometimes supposed. In most contexts, it is usually better to fight environmental battles through lawful tactics such as rallies, marches, boycotts, lawsuits, lobbying, and the like. As the world—far too slowly—begins to take the climate crisis seriously, environmental activism is likely to become more frequent and probably more militant. Most climate activism, one hopes, will be passionate but peaceful and law-abiding. But civil disobedience may also play a role in the battle for climate justice and a livable earth. For as history shows, it can be a powerful tool in the fight for a better world.

Chapter Summary

1. Environmental protests may be legal or illegal. In democratic societies common legal means of environmental protest include peaceful rallies

15. Foreman, *Confessions of an Eco-Warrior*, pp. 3, 26.
16. Cohen, *Civil Disobedience*, p. 114.
17. For examples, see John Hart Ely, *Democracy and Distrust: A Theory of Judicial Review* (Cambridge, MA: Harvard University Press, 1980), pp. 50–52.

and marches, speeches, boycotts, petitions, and letter-writing campaigns. Illegal forms of protest include things such as sit-ins, unlawful occupations, roadblocks, tree-spiking, ecosabotage, vandalism, fire-bombings, and other criminal acts of trespassing, property destruction, or violence.

2. There are four major forms of conscientious lawbreaking: conscientious refusal, conscientious evasion, militant action, and civil disobedience. Conscientious refusal is a public refusal to obey a law or lawful command for personal ethical or religious reasons (e.g., because of a personal conviction that all wars are immoral). Conscientious evasion is an attempt to evade the law and avoid notice or capture for personal ethical or religious reasons (e.g., secretly aiding undocumented migrants for faith-based reasons). Militant action involves conscientiously motivated acts of violence, coercion, or property-destruction. Civil disobedience is a public, nonviolent, deliberately unlawful act of protest, usually done with the aim of bringing about a change in the law or the policies of a government. Of these four forms of conscientious lawbreaking, civil disobedience is usually the easiest to defend.

3. Various factors must be weighed in deciding whether a particular act of environmental civil disobedience is ethically justified. Among these are consequentialist considerations such as the seriousness of the threatened environmental harm, the risk of violence, the likelihood of serious backlash, and the likelihood that the protest will actually succeed. Various duty-centered principles and virtue-centered considerations may also be relevant to the moral equation. When all pertinent factors are duly weighed, relatively few acts of ecological civil disobedience are likely to be justified (at least in democratic, well-ordered, and substantially just societies). Things may change, however, as the fight against climate change becomes more critical.

Discussion Questions

1. What are some common legal forms of environmental protest? How do conscientious refusal, conscientious evasion, militant action, and civil disobedience differ from one another?

2. Do you agree that it is sometimes morally permissible to disobey a law? If so, give some examples.

3. Why is civil disobedience generally easier to justify than conscientiously motivated militant action?
4. Is militant action ever a justifiable form of environmental protest? If so, under what circumstances?
5. Is civil disobedience ever a justifiable form of environmental protest? If so, give some examples.
6. Can civil disobedience as a form of environmental protest ever be justified by appeals to "higher law"? Why or why not?
7. Is civil disobedience a legitimate tactic in the battle against climate change? Defend your view.

Further Reading

Classic readings on the obligation to obey the law include Sophocles's *Antigone*, Plato's *Crito,* and Henry David Thoreau's "Civil Disobedience" (all available online); Mahatma Gandhi, "Non-violence" and "Civil Disobedience," reprinted in Dennis Dalton, ed., *Mahatma Gandhi: Selected Political Writings* (Indianapolis, IN: Hackett, 1996); and Martin Luther King Jr., "Letter from Birmingham City Jail" (available online). Edward Abbey's novel, *The Monkey Wrench Gang* (New York: Avon Books, 1976), was a major inspiration for many early eco-saboteurs. For an older but still useful reader on radical ecological activism, see Peter C. List, ed., *Radical Environmentalism: Philosophy and Tactics* (Belmont, CA: Wadsworth, 1993), which includes key readings on militant environmental groups such as Greenpeace, The Sea Shepherd Society, and Earth First! For informative accounts of radical environmental movements in the 1970s and 1980s, see Christopher Manes, *Green Rage* (Boston, MA: Little, Brown, 1990); Rik Scarce, *Eco-Warriors: Understanding the Radical Environmental Movement* (Chicago, IL: The Noble Press, 1990); Susan Zakin, *Coyotes and Town Dogs: Earth First! and the Environmental Movement* (New York: Viking, 1993); and Dave Foreman, *Confessions of an Eco-Warrior* (New York: Harmony Books, 1991). For a useful collection of readings on civil disobedience, see Hugo Bedau, ed., *Civil Disobedience in Focus* (New York: Routledge, 1991).

Afterword

Looking to the Future

Since its emergence as an academic discipline in the early 1970s, environmental ethics has come a long way in a short amount of time. Five decades ago there were no classes, textbooks, conferences, or journals in environmental ethics. Today the field is flourishing. Now, for the first time in Western history, thinkers are grappling in a serious and sustained way with ethical concerns that arise from our interactions with nature and the nonhuman fellow life-forms with whom we share this planet.

Why did it take so long for environmental ethics to emerge? Partly, no doubt, because of the long dominant mindset of what Pope Francis has called "tyrannical anthropocentricism."[1] Until quite recently, most Westerners believed the view that only human beings are worthy of moral respect and concern. On such a view, humans may treat nature and nonhuman organisms pretty much as they please, provided no human rights or interests are wrongfully infringed. Only when someone or something is seen as "morally considerable" do ethical concerns arise.

The decade in which environmental ethics arose, the 1970s, was a time of great intellectual ferment. Many so-called liberation movements—most notably, black liberation, women's liberation, and gay liberation—made significant progress in this era. Early advocates of "animal liberation," like Peter Singer, were part of that great social and ethical paradigm shift. All the various liberation movements were egalitarian, anti-discrimination campaigns that rejected traditional hierarchical ways of thinking. Environmental ethicists that reject anthropocentrism and see nature as having value for its own sake are motivated by a similar spirit.

In the few decades since it arose, environmental ethics has made a great deal of progress. Much useful and important work, however, remains to be done. Let's briefly examine four areas where laborers in the ecological vineyard are especially needed.

1. Pope Francis, *Laudato Si'*, May 24, 2015, §68 (online). Web. 26 April 2020.

In discussing biocentrism—the view that all nonhuman organisms have considerable moral standing—we noted the importance of developing what Paul Taylor calls "priority rules" for handling conflicts between human interests and those of nonhuman life-forms. Without such rules—or at least general guidelines—we have no principled and coherent way to decide when nature's good rightly supersedes our own. Lacking such rules, we will constantly find ourselves yielding to temptation and engaging in conduct that, on reflection, we recognize to be arrogant or unjust. As we saw, Taylor himself proposed a number of priority rules, but these, we noted, are problematic for various reasons. Consequently, one important question for environmental ethicists to tackle is: What would a satisfactory set of human/nonhuman priority rules or general principles be like?

Another pressing issue relates to ecocentrism, the view that our focal environmental concern should be for the good of species, ecosystems, or other ecological wholes, rather than for the good of individual organisms. In our discussion of ecocentrism in Chapters 5 and 7, we noted that ecocentrism is now by far the most popular environmental ethic. We also saw, however, that we don't currently have a well-developed and critically defensible ecocentric theory. Most contemporary versions of ecocentric ethics are rooted in Aldo Leopold's influential land ethic, which we noted raises many problems. In Chapter 7 I sketched a version of ecocentric ethics I called "moderate ecocentrism." Such a view, I suggested, seems to be the best general framework for environmental ethics. But the theory I proposed was just a sketch, or proto-theory. Much work would be needed to flesh it out and defend its central claims.

A third area of environmental concern in which ethical theorizing is badly needed is biodiversity preservation and recovery. As we saw earlier, a recent landmark United Nations study found that one-eighth of all species of plants and animals are threatened with extinction, many in a few decades.[2] No matter what we do, some species will be lost as a result of climate change and other anthropogenic causes. Thus, many hard and painful questions must be asked about which species we should try hardest to save, and at what cost and effort. Tragically, such hard choices simply cannot be avoided.

Finally, bold and creative philosophical thinking is urgently needed on climate ethics. Stephen Gardiner aptly calls climate change a "perfect moral storm"[3] because of the unprecedented ethical, political, and psychological

2. Isabelle Gerretsen, "One Million Species Threatened with Extinction Because of Humans," *CNN*, May 7, 2019. Web. 26 April 2020.

3. Stephen M. Gardiner, *A Perfect Moral Storm: The Ethical Tragedy of Climate Change* (New York: Oxford University Press, 2011).

challenges it poses to an effective response. As Gardiner convincingly argues, climate change presents a complex, long-term, international, intergenerational collective action problem that current political, economic, and ethical systems are poorly equipped to handle. Human ethical instincts and responses did not evolve to respond effectively to the kinds of remote, long-term, uncertain, and often unwitting harms caused by climate change. In grappling with the ethical dimensions of the climate crisis, philosophers will need to develop robust new theories of intergenerational ethics, global justice, global politics, and interspecies ethics. Opportunities for fresh and impactful philosophical work in this field abound.

In addressing such issues, philosophers will be able to build on the impressive body of work environmental ethicists have achieved over the past few decades. As we labor to heal our sick planet, environmental philosophy must play a vital role.

Glossary

animal behavior argument: An argument that asserts that animals have no moral rights because they don't respect the "rights" of other animals.

animism: The belief that many or all natural objects, as well as all living things, have a mind or soul.

anthropocentrism: A strongly human-centered view of nature. Commonly, anthropocentrists claim that only human beings have inherent value or that only humans have moral standing.

biocentric egalitarianism: A form of biocentrism that holds that all living things have equal inherent value and equal moral standing.

biocentric outlook on nature: A set of beliefs and attitudes respecting nature endorsed by the environmental philosopher Paul Taylor. The biocentric outlook rejects human superiority, sees humans as coequal members of earth's community of life, recognizes the interdependence of all life on earth, and understands that all living things exhibit goal-like behavior, pursuing their own good in their own unique way.

biocentrism: A life-centered view of nature. Biocentrists claim that all living things, plants as well as animals, have substantive moral standing.

biodiversity: The number and varieties of life-forms that exist in the world or in a particular region or ecosystem.

biophilia hypothesis: The idea, first popularized by biologist E. O. Wilson, that humans have an innate love of nature and naturally seek and enjoy communion with nonhuman life-forms.

categorical imperative: The supreme principle of morality, according to Immanuel Kant. One formulation of the categorical imperative asserts that one should only act on "maxims" (rules) that one would wish to become universal rules of human conduct. Another asserts that one should always treat oneself and other humans as ends in themselves, never as mere means.

civil disobedience: A nonviolent, deliberately unlawful act of protest, usually done with the aim of bringing about a change in the law or the policies of the government.

climate change: Changes in global or regional climate patterns (especially warmer temperatures) occurring after the Industrial Revolution, attributed to increased levels of carbon dioxide and other greenhouse gases in the atmosphere, due mainly to burning fossil fuels.

conscientious evasion: A covert attempt to evade a law or lawful command, motivated by personal ethical or religious beliefs.

conscientious refusal: A refusal to comply with a law or lawful command, motivated by personal ethical or religious beliefs.

consumption: In economic terms, the buying or using of goods and services in the economy, usually in a way that reduces those goods or services, or makes them unavailable to others.

contractarianism: A moral theory that holds that ethical norms arise from mutual agreements or social contracts. The contractarian argument states that animals cannot have moral rights or moral standing because they cannot enter contracts or agreements.

deep ecology: A radical environmental ethic that stresses the need for deep, far-reaching, and fundamental changes in the way humans live and interact with nature. The core tenets of deep ecology are commonly identified with the eight-point Deep Ecology Platform formulated by Arne Naess and George Sessions.

demographic transition: A widely observed effect in which both birth and death rates drop when societies become more developed.

descriptive ethics: The empirical study of moral standards and moral language. Descriptive ethics attempts to explain or describe what ethical norms are actually embraced, not to prescribe which norms should be embraced.

different-people argument (or the non-identity argument or the disappearing beneficiaries argument): An argument put forward by David Parfit that claims that future generations have no moral rights, and present generations have no moral duties to them, because any major action we take today will result in different people being born in the future. Thus, no matter what big actions we take today, future generations will not be harmed or made worse off, because they wouldn't even have existed if we had acted otherwise.

dominion argument: The argument that animals have no moral rights or moral standing because God has given humans dominion over them.

duty ethics (or **deontological ethics or Kantian ethics**): A type of moral theory that focuses primarily on ethical duties and rules, rather than on the consequences of actions or the characters of moral agents.

ecocentrism (or **ecological holism**): The view that our primary environmental concern should be for the well-being of species, ecosystems, or other ecological wholes, rather than for the well-being of individual plants and animals. The best-known proponent of ecocentrism is Aldo Leopold.

ecofeminism: An environmental theory that opposes sexism (the systematic oppression of women) and naturism (the systematic oppression of nature), and asserts that there are important connections between both forms of oppression.

ecology: The science that studies living things and their interactions in relationship to their environments.

ecosabotage (or **ecotage** or **monkey-wrenching**): Unlawfully damaging or destroying property (usually covertly) for the sake of some environmental cause.

environment: The conditions or surroundings (especially natural conditions or surroundings) that an organism lives in or occupies.

environmental ethics: The branch of philosophy that studies how humans should view and treat nature and the environment.

environmental fascism: A common objection to ecocentric environmental theories. Like political fascism, which wrongly subordinates the individual to the larger social good, ecocentrism is alleged to wrongly subordinate the good of individual organisms to the good of larger wholes or collective entities, such as species or ecosystems.

environmental justice: A fair and equitable distribution of environmental benefits and burdens.

environmental racism: A form of discrimination that inflicts disproportional environmental burdens or harms on members of racial or ethnic minorities.

environmental stewardship: Acting as a faithful trustee or responsible caretaker of earth's natural beauties and resources.

ethical egoism: A moral theory that asserts that everyone ought always to act in his or her own long-term self-interest.

ethical extensionism: The view that moral rights or moral standing ought to be extended to animals, plants, ecosystems, and other objects in nature that traditionally have not been viewed as deserving of moral status or consideration.

ethics: The branch of philosophy concerned with morality. In a descriptive sense, ethics refers to ethical values that are actually accepted by an individual, a group, or a society. In a normative sense, ethics seeks to identify ethical values that *ought* to be accepted.

eudaimonist virtue ethics: A form of virtue ethics—first systematically defended by Aristotle—that holds that the proper goal or final end of human existence is happiness or flourishing (*eudaimonia*, in Greek).

genetically modified foods: Foods whose genetic material (DNA) has been altered by means of genetic engineering.

hedonistic utilitarianism: A form of utilitarianism that asserts that an act is morally right just in case it maximizes net happiness or pleasure, that is, produces the greatest happiness or pleasure for the greatest number.

ignorance argument: The argument that future generations have no moral rights, or no moral standing, because we can know almost nothing about what they might want or need.

inherent value. *See* intrinsic value.

intrinsic value: The value a thing has in itself or for its own sake, apart from its value as a means or conduit to other goods or values.

land ethic: A type of ecocentric environmental ethic proposed by Aldo Leopold. The land ethic holds that moral status should be extended to animals, plants, soils, waters, or collectively: the land. The basic norm ("summary moral maxim") of the land ethic is: "A thing is right when it tends to preserve the integrity, stability, and beauty of the biotic community. It is wrong when it tends otherwise."

line-drawing argument: The claim that animals have no moral rights or moral standing because it is impossible to draw clear lines between animals that deserve moral consideration and those that do not.

locavore: A person who only or mainly eats locally grown and produced foods.

logic of domination: A form of thinking that claims that superiors have a right to dominate or oppress inferiors. According to ecofeminist Karen Warren, both sexism (the systematic subordination of women) and naturism (the systematic subordination of nature) are rooted in a discriminatory logic of domination.

metaethics: The branch of ethics that addresses "second-order" questions of ethics, such as the meaning of moral language or the nature of moral values, rather than first-order moral questions, such as what sorts of acts are morally right or wrong.

militant action: A form of illegal protest or direct action that involves violence, coercion, or destruction of property.

moderate ecocentrism: An environmental ethic that holds that all living things have inherent worth; that some organisms have greater inherent worth than others; that humans have greater inherent worth than any other organism; and that our primary environmental concern should ordinarily—but not always—be for the good of species and other ecological wholes, rather than for the good of individual organisms.

moral relativism (or **cultural moral relativism**): A moral theory that asserts that an act is morally right for a person A just in case A's culture or society believes that the act is morally right.

moral standing (or **moral considerability**): The intrinsic value or inherent worth something has that makes it deserving of ethical consideration, respect, and concern.

moral subjectivism: A moral theory that holds that an act is morally right for a person A just in case A believes that the act is morally right.

negative duty: An ethical duty to refrain from performing a prohibited action, such as theft, in contrast to a positive duty, which requires that some conduct be actively engaged in, such as rescuing a stranded motorist.

no-claim argument: The argument that animals have no moral rights or moral standing because animals cannot assert claims to such rights or standing.

normative ethics: That branch of ethics that attempts to answer value questions, such as what sorts of actions are morally right or wrong, or what qualities of character one should try to develop.

philosophy: The attempt to answer deep and fundamental questions of meaning, value, knowing the ultimate nature of reality, and other big questions that cannot be answered by science or observation by the senses.

pluralistic duty ethics: A form of duty ethics that holds that there are several basic moral principles, not just one fundamental principle, as Kant claimed. The Oxford philosopher W. D. Ross was a leading proponent of pluralistic duty ethics.

pluralistic utilitarianism: A form of utilitarianism that asserts that there are several basic intrinsic goods (e.g., knowledge, moral goodness, beauty, and friendship), not just one, such as happiness.

positive duty: An affirmative ethical duty to perform an act of some kind, such as to relieve a person in distress, in contrast to a negative duty, which is a duty to refrain from some kind of prohibited action.

precautionary principle: A widely accepted principle of risk management. In one popular formulation, it requires that precautionary measures be taken whenever an activity raises threats of harm to human health or the environment, even if some cause and effect relationships are not fully established scientifically.

preference utilitarianism: A form of utilitarian ethics that holds that an act is morally right just in case it maximizes net preference-satisfaction. The animal-welfare advocate Peter Singer is a proponent of preference utilitarianism.

prima facie duties: W. D. Ross's term for moral duties that are not "absolute" or exceptionless, but rather apply "at first glance" and may in some contexts be overridden by weightier duties.

principle of dignity: One of Kant's two main formulations of his supreme principle of morality, the categorical imperative. The principle of dignity asserts that one should always treat humanity, whether in oneself or in others, as an end in itself, and never as a mere means.

principle of equal consideration of interests: The ethical principle, endorsed by Peter Singer, that equal interests should be treated equally (unless there are compelling reasons to treat them differently). According to Singer, this is the true basis of moral equality and the ultimate reason why a bias in favor of one's species ("speciesism") is wrong.

principle of universal law: One of Kant's two main formulations of his supreme principle of morality, the categorical imperative. The principle of universal law asserts that one should act only on those maxims (that is, rules or principles) that one would wish to see become universal rules of human conduct.

reverence for life: Albert Schweitzer's supreme principle of morality. According to Schweitzer, all forms of life are sacred and should be treated with reverence.

rights view: Tom Regan's ethical theory that holds that all animals that are "subjects of a life" (that is, roughly, sentient beings that have desires, purposes, etc.) have equal inherent value and are entitled to certain basic moral rights.

sentientism: The view that only sentient animals have moral rights or moral standing.

speciesism: An unjustifiable and discriminatory bias in favor of one's own species.

standpoint argument: An argument offered by Paul Taylor to show that humans are not superior in inherent value to other life-forms. According to the standpoint argument, all alleged "superior" human qualities, such as rationality and moral discernment, are superior qualities only from a human point of view, and thus not objectively superior at all.

subjects of a life: Tom Regan's term for animals that have a high enough level of consciousness to possess moral standing and equal inherent worth. As Regan uses the term, subjects of a life are "higher" or more complex sentient animals (e.g., dogs, cows, and chickens) that have desires, purposes, emotions—in short, conscious experiences and a life that matters to them.

summary moral maxim: The basic moral principle of Aldo Leopold's ecocentric land ethic. The summary moral maxim states: A thing is right when it tends to preserve the integrity, stability, and beauty of a biotic community. It is wrong when it tends otherwise.

supererogatory acts: Acts that go above and beyond the call of moral duty; that is, highly meritorious acts that go beyond what is strictly required.

target-centered (or "pluralistic") virtue ethics: A type of virtue ethics that conceives of virtues as traits or excellences of mind or character that are conducive to the achievement of basic goods, such as freedom, justice, beauty, or knowledge.

techno-fix: A significant or game-changing technological solution to climate change.

teleological centers of life: Paul Taylor's term for the goal-like tendencies of plants and animals to pursue their own good in their own way. According to Taylor, all living things are teleological centers of life.

utilitarianism: A moral theory that asserts that an act is morally right just in case it maximizes net happiness or welfare.

virtue: A good habit, disposition, or quality of mind or character (e.g., kindness or honesty)

virtue ethics: A type of ethical theory that places primary emphasis on questions of character and virtue rather than on consequences, duties, or rules, as other leading ethical theories do.

wilderness: A tract or region uncultivated and uninhabited by humans. In the American environmental context, "wilderness" can refer either to public lands that have been officially set aside as protected wilderness (de jure wilderness) or to public lands that meet the federal definition of wilderness but have not as yet been set aside as such (de facto wilderness). According to some critics of wilderness or wilderness protection, there is also a "received view" of wilderness that conceives of it as "pristine nature" or "virgin land" that has not been impacted by humans in any significant way.

Index